A HANDBOOK TO THE

UNIVERSE

Explorations of Matter, Energy, Space, and Time for Beginning Scientific Thinkers

RICHARD PAUL

CHICAGO REVIEW PRESS

Library of Congress Cataloging-in-Publication Data

Paul, Richard, 1948–
    A handbook to the universe : explorations of matter, energy, space, and time for beginning scientific thinkers / Richard Paul.—1st ed.
      p. cm.
    A Ziggurat book.
    Includes bibliographical references and index.
    Summary: Explains microwaves, sonic booms, time zones, tides, rainbows, and many other topics. Includes experiments.
    ISBN 1-55652-172-3 : $14.95
    1. Physical sciences—Juvenile literature.   2. Physics—Juvenile literature.   3. Astronomy—Juvenile literature. [1. Physical sciences. 2. Physics. 3. Astronomy.]   I. Title.
Q163.P2929   1993
500.2—dc20
                                               92-39670
                                               CIP
                                               AC

Interior Cartoons © Michael Fleishman

Published by Chicago Review Press, Incorporated
814 North Franklin Street
Chicago, Illinois 60610

Printed in the United States of America

5 4 3 2 1

For Kris,

who kept my feet on the ground
when my head was in the clouds

"Out yonder there was this huge world, which exists independently of us human beings and stands before us like a great eternal riddle. . . . The contemplation of this world beckoned like a liberation. . . ."
—Albert Einstein 1879–1955

CONTENTS

PREFACE   xv

I. MATTER   1

I.1. A Material World
A definition of matter; ancient attitudes toward matter; the
four Greek elements.

I.2. A Primitive Theory
Democritus proposes atoms; Aristotle rejects atoms and
develops the first comprehensive theory of matter.

I.3. A Suffocating Status Quo
Advent of Dark Ages halts progress toward understanding
matter; Aristotelian ideas linger; alchemy; Galileo refutes
Aristotle.

I.4. Documented Facts
Boyle resurrects the idea of atoms and revises the idea of
elements; phlogiston invented to explain combustion.

I.5. Fundamental Types of Matter
Lavoisier properly explains combustion; Lavoisier discovers
law of conservation of matter and identifies chemical
elements.

I.6. A Hidden Mechanism
Dalton deduces the existence of atoms from the behavior of
elements as they participate in chemical reactions; Dalton
assigns weights to atoms and suggests means of atomic
interaction.

I.7. An Underlying Organization
Numerous elements discovered and coincidences among
them noticed; Mendeleev devises the periodic chart.

I.8. A Full Deck
The number and nature of chemical elements; their
occurrence, importance, and use; allotropes, the multiple
forms of a single element (carbon as both graphite and
diamond).

I.9. Complex Substances
Chemical reactions among elements; the creation of molecules from atoms and entirely new, complex substances from basic, elemental ingredients.

I.10. A Jumble of Particles
The coexistence of various elements and compounds in physical mixtures; the differences between physical mixtures and chemical compounds.

I.11. Familiar Mixtures
Matter commonly encountered as mixtures of many ingredients; the retrieval of these ingredients by physical and chemical means; mining metals as a typical example.

I.12. Analysis and Synthesis
Chemistry's use in reassembling retrieved ingredients in new and useful ways; refining petroleum as a typical example.

I.13. Two Universal Properties
Matter's value as a function of its properties; the universal properties of mass and volume; mass versus weight.

I.14. Density
Density as a specific property that can be used to identify types of matter; density versus weight; density's importance.

I.15. A Multitude of Characteristics
The nature of other specific properties such as heat capacity, hardness, solubility, and malleability and how these properties serve as our basis for dealing with matter.

I.16. Submicroscopic Goings-on
How the kinetic molecular theory of matter explains the operation of specific properties like density and hardness.

I.17. Three Peculiar Personalities
A description of matter's gaseous, liquid, and solid states in molecular terms; water as the ideal example.

I.18. Solid Stuff
The nature of solids; a discussion of amorphous-type solids like glass and plastics; the history and variety of plastics.

I.19. Crystalline Structure
A discussion of crystalline-type solids like gems and minerals; their structure and general behaviors.

I.20. A Metallic Medley
Metals and other important crystalline solids; properties of elemental metals and of metallic alloys like steel and brass.

I.21. Capricious Liquids
The nature of liquids; different types of liquids and their distinguishing characteristics.

I.22. Wet, Wonderful Water
Water's uniqueness, both in terms of quantity and quality; its global functions and peculiarities; its use as a standard.

I.23. Elusive Gases
Matter's gaseous state; the universal response of all gases to changes in volume, temperature, and pressure.

I.24. An Ocean of Air
Galileo determines that air has weight; atmospheric pressure and how it creates weather; how a plane flies.

I.25. Temporary Compromises
Gas/liquid/solid combinations existing as solutions and as colloids; common examples of these combinations.

I.26. Subatomic Fragments
Discovery of the electron and proton; atomic radiation and radioactive decay among the elements.

I.27. A Miniature Solar System
Exploring the design and operation of the atom; atomic structure, number, and weight; atomic structure as the determinant of chemical behavior.

I.28. Nuclear Physics
How the workings of the atom explain carbon-14 dating and the laser; the creation of new types of atoms called isotopes and how they're used; creating new elements.

II. ENERGY

71

II.1. The Agent of Change
The constant changes undergone by matter; energy as the cause of these changes; early inability to distinguish between matter and energy; Thompson's discovery of energy.

II.2. The Conservation of Energy
The different forms of energy; how energy remains constant in amount even as it changes forms.

II.3. Transformations of Energy
Energy's kinetic and potential states; examples of changes from one state to the other; the involvement of heat in these changes.

II.4. Mechanics

Mechanics as the most obvious form of energy; the early triumphs of Archimedes and the early failures of Aristotle; medieval confusion about mechanics.

II.5. Inertia

Galileo puts mechanics on a solid theoretical basis; Galileo's work with falling bodies and his formulation of inertia.

II.6. Three Laws of Motion

Newton's three laws of motion stated, explained, and illustrated with common examples.

II.7. The Conservation of Momentum

The concept of momentum and its relationship to mass and velocity; the conservation of momentum in typical events; the operation of rockets, jets, and similar phenomena.

II.8. Friction

The occurrence and nature of friction; its functions, both good and bad; its relationship to heat.

II.9. The Nature of Heat

Heat described in molecular terms; coldness versus heat; heat's involvement in chemical reactions; combustion.

II.10. Temperature, Not Heat

The difference between temperature and heat; expansion due to temperature change; the operation of thermometers; various temperature scales.

II.11. The Latent Heat of Fusion

Heat's role as matter changes back and forth between its liquid and solid states; demonstrations using water and ice.

II.12. The Latent Heat of Vaporization

Heat's role as matter changes back and forth between its gaseous and liquid states; the operation of a refrigerator; pressure and its effect on vaporization; pressure cookers.

II.13. Evaporation

Evaporation described in molecular terms; factors that aid or hinder evaporation; perspiration as a cooling mechanism.

II.14. Weather Events

Humidity and relative humidity; evaporation, condensation, humidity, and temperature and how they create clouds, fog, rain, snow, dew, and frost.

II.15. The Flow of Heat
The one-way flow of heat from hot to cold objects via conduction, convection, and radiation.

II.16. Sound
Sound as a form of energy; a description of sound waves; the nature and behavior of sound waves.

II.17. Acoustics
Sources of sound; conductors and inhibitors of sound; reverberations and echoes.

II.18. The Speed of Sound
What determines the speed of sound; the sound barrier; breaking the sound barrier; sonic booms.

II.19. The Subtleties of Hearing
The capabilities of the human ear; binaural hearing; the Doppler effect.

II.20. Vision
Light as a form of energy; sources of light; uses of light; how we use light to see; depth perception.

II.21. The Behavior of Light
The propagation of light through transparent media and through a vacuum; the speed of light in various media; refraction's role in mirages, optics, and twinkling stars.

II.22. Light Waves
The physical nature of light; the relationship between frequency and wavelength; brightness.

II.23. The Perception of Color
Newton's discovery of the cause of color; white versus colored light; prisms and rainbows.

II.24. Reflection
Mirrors; white, colored, and black objects; the Tyndall effect and the appearance of shadows, clouds, and the sky.

II.25. Electromagnetic Radiation
Light as a special type of electromagnetic radiation; radar; microwaves in radio and television transmission and in ovens; ultraviolet radiation and sunburn; x-rays.

II.26. The Magic of Magnetism
Electromagnetic radiation as the interplay between electrical and magnetic forces; magnetism's discovery and early uses; the laws of magnetism; how a compass works; the northern lights.

II.27. Static Electricity
An atomic description of electricity; electrical charges; insulators and conductors; static electricity; lightning.

II.28. Electric Current
Volta's invention of the battery; the characteristics of an electrical current; how batteries and other electrical devices work.

II.29. Nuclear Energy
The discovery of radioactive decay; the mechanics of atomic disintegration; the interconvertibility of mass and energy.

II.30. A Chain Reaction
The search for practical nuclear energy; the critical role of uranium; the first nuclear reactor.

II.31. The Atomic Bomb
Creating the atomic bomb; the source of the A-bomb's power; the operation of modern nuclear reactors; radioactive waste.

II.32. Nuclear Fusion
Fusion versus fission; how fusion operates; the desirability and difficulty of controlling nuclear fusion.

III. SPACE   155

III.1. The Endless Expanse of Space
Early assumptions that the heavens were of a separate reality; the seven travelers and the special status accorded them.

III.2. Logic and Geometry
The Greeks explain lunar and solar eclipses, determine the world to be round, and measure its size using logic and geometry.

III.3. The Greek Universe
Aristotle's theoretical two-part universe; problems presented by the planets; the epicycle as a solution; geocentricity assumed; Ptolemy's final draft.

III.4. A Revolutionary Idea
Practical demands made of the sky finally make Ptolemy obsolete; Copernicus proposes a new, heliocentric alternative.

III.5. A Simple, Consistent Design
Brahe assembles the evidence by which the design of the

universe is to be judged; Kepler succeeds in explaining planetary motion in terms of ellipses.

III.6. A Spokesman for the New Order
Traditional Greek geocentricity versus revolutionary Copernican heliocentricity; the telescope invented; Galileo's crusade for heliocentricity; the trial of Galileo.

III.7. The Final Triumph
The lack of a believable motivator of celestial events; Newton's explanation via gravity; a truly unified universe.

III.8. Universal Gravitation
The *Principia;* the nature of gravity; the operation of the solar system in gravitational terms.

III.9. Celestial Mechanics
Newton's three laws of motion as applied to heavenly bodies; the operation of the universe in terms of gravity and inertia; orbits.

III.10. The Earth in Orbit
A description of planet Earth and its path about the sun; two proofs of the earth's motion: the aberration of starlight and stellar parallax.

III.11. The Rotation of the Earth
A description of the earth's rotation on its axis; Foucault's pendulum as proof of earthly rotation; how rotation effects events on the face of the globe; the Coriolis effect.

III.12. The Seasons of the Year
The tilt of the earth's axis; the solstices and equinoxes; the distribution of sunlight over the face of the globe during its orbit; why the equator is hot and the poles are cold.

III.13. The Moon
The size and location of the moon; the illumination of the moon; the view of the moon from the earth.

III.14. The Lunar Orbit
How the moon's orbit accounts for the variations in moonrise and moonset times; how the moon's orbit accounts for its phases; lunar and solar eclipses; the tides.

III.15. The Solar System
The arrangement of the sun and the planets; the ecliptic; the common motion of most solar system members.

III.16. The Fixed Stars
The unchanging pattern of stars; galaxies and the Milky Way; apparent motion of the nighttime sky; the zodiac.

III.17. The Universe
The objects of outer space; collecting and analyzing starlight; the expanding universe; three cosmic theories.

IV. TIME
201

IV.1. The Essence of Time
Universal changes; the relationship of time and change; time defined in terms of change.

IV.2. Older, Never Younger
Time interval and time epoch; the reversibility of many natural events; the irreversibility of time.

IV.3. The Past and the Future
The first human concepts of the extent of time; civilization and the expansion of time's boundaries; Christianity's demarcation of time's beginning and end.

IV.4. The Investigation of Time
Leclerc challenges the traditional date of time's beginning using the scientific method; uniformitarianism; dating with isotopic half-lives; the age of the earth.

IV.5. The Beginning of Time
Theories about the origin of the earth and the solar system; theories about the origin of the universe; the age of the universe.

IV.6. Three Natural Clocks
Measuring current time; the year, the month, and the day as natural units of time; their basic incompatibility.

IV.7. The Lunar Calendar
The need to keep track of time; the earliest calendars and their deficiencies; intercalation.

IV.8. The Julian Calendar
The switch from lunar to solar calendars; the Roman origin of our present calendar.

IV.9. The Numbering of Years
The formalization of the week; various numbering schemes for the years; the adoption of the current B.C./A.D. system.

IV.10. The Gregorian Calendar
The shortcomings of the Julian calendar; the Gregorian refinements; complications arising from the shift to the

Gregorian calendar; worldwide acceptance of the Gregorian calendar.

IV.11. Timekeeping
The day as the standard of current time; dividing the day into smaller units; early clocks.

IV.12. Supreme Accuracy
Mechanical clocks; the evolution of clocks relying on natural periodic motions.

IV.13. Irreconcilable Standards
The second as a fraction of a day; the second as a multiple of other natural rhythms; the dilemma of having two different seconds and its resolution.

IV.14. Standard Time
The relative nature of time; the use of the sun as a reference point; the need for synchronization; standard time established.

IV.15. Time Zones
An explanation of why time zones are needed and how they work; jet lag; the International Date Line.

IV.16. The End of Time
The likelihood of time's end; the various ways time could cease; the probable date of time's demise.

V. REALITY   243

V.1. A Confusing Predicament
The human urge to explore and understand reality; early efforts and accomplishments; the conceptualization of nature as deities.

V.2. Natural Philosophy
An accumulation of data reveals patterns in nature; the Greeks trade idols for intellect; the world according to Aristotle.

V.3. The Birth of Science
Copernicus and Galileo undo the universe of Aristotle; Newton establishes classical physics; the extreme scientific attitude of determinism.

V.4. A Fatal Flaw
Early resistance to science; the emotional appeal of ether; the requirement of ether in a Newtonian universe.

V.5. The Definition of Motion
The need for a static reference point from which to measure motion; the use of the earth, the sun, and then ether as that reference point.

V.6. Junking Classical Physics
The search for ether and the failure to find it; the ether dilemma; Einstein's daring solution.

V.7. The Theory of Relativity
The assumptions upon which Einstein's universe is based; a description of a new reality based upon those assumptions.

V.8. Mickey and Minnie
How relativity affects our perception of reality; an example demonstrating the variable nature of time, space, and velocity.

V.9. Confirming Relativity
Evidence supporting Einstein's radical ideas about gravity, space, energy, time, and mass; the apparent superiority of the relativistic model of reality.

V.10. Curved Space
The modern concept of space as a real physical entity subject to distortions and irregularities; its finite yet unbounded nature.

V.11. Quantum Mechanics
The long-standing debate over the nature of light; the discovery of quanta and the fact that energy, like matter, is discontinuous.

V.12. The Uncertainty Principle
The difference between classical and quantum mechanics; the replacement of certainty with probability; the theoretical limits of knowledge.

V.13. A Matter of Opinion
The accidental evolution of current reality; our best efforts to explain it; the success and ultimate failure of science in solving the riddle of reality.

TIMELINE   273

SELECTED BIBLIOGRAPHY   277

INDEX   279

Preface

So, here we are, stuck on a relatively small clump of dirt and water, aimlessly circling a minor star somewhere out in the cosmic boondocks. It's a bit of a predicament, to say the least, and not one that offers us a whole lot of options. As a matter of fact, there's really just one course of action available to us under the circumstances. Following the dictates of nature, we can only do what we have to do. We struggle to survive as best we can.

Like the birds, the bees, and rest of the beasts, we wage this life-or-death battle with every weapon at our disposal. In our case more often than not it's our brains we call upon to sustain us. There's a good reason for this, of course. Our species, as you probably know, is neither especially swift, strong, hardy, nor otherwise well equipped to endure the hardships provided by planet Earth. It is, however, supremely intelligent. We are, after all, *Homo sapiens*—thinking human beings.

Our strategy apparently works. Or at least it has so far. By relying on little more than what resides between our ears we've been able not only to survive, but prosper. Every day there are more of us, living longer and on the whole better, than there were the day before. In this we are unique. Out of thousands of surviving species ours is the only one that not only tries but sometimes succeeds in improving its lot.

But what is it that makes a gelatinous glob of gray matter more powerful than a thick hide, venomous fangs, or sharp claws in the battle against a harsh environment? How, exactly, does the possession of an oversized cerebrum contribute to proliferation, longevity, and comfort? And how does the ability to think great thoughts enable an otherwise somewhat puny species to transform itself from a rag-tag band of brutish cave dwellers into a powerful army of movers and shakers?

What follows, in case you hadn't already guessed, is an attempt to answer these questions. It's an examination of how we've used our wits to cope with our situation here on planet Earth. It's an account of how we've methodically studied nature, learned the laws by which it operates, and then used that knowledge to gain a measure of control over our environment and our lifestyle. Finally, it's a summary of the most important of those laws and a brief description of how the natural world works.

That's right, it's a science book. But wait, let me explain.

What you hold in your hands is unlike any other science book you've ever read, tried to read, or were supposed to have read. This one, although certain to be useful to anyone interested in science, is also for those of you who aren't particularly fond of, "good at," or even comfortable with it. This is for those of you who have always wanted to understand how the physical world works but, for various reasons, have never succeeded. Here, in a format certain to overcome your anxieties, is everything you've always wanted to know about microwaves, sonic booms, time zones, molecules, nuclear power, the tides, ozone, plastic, clouds, electricity, static cling, relative humidity, bathtub rings, rainbows, and a thousand or so other fascinating topics, but were afraid to ask.

The science presented here differs from what you've encountered elsewhere in several important ways. In the first place, it's risk free. This isn't a textbook, at least not in any traditional sense. There are no chapter reviews here, no assignments, exercises, or tests. The idea is not to find out how well you can memorize facts, but rather to help you satisfy your curiosity about how the world operates.

Besides being presented in a practical rather than academic way, the material included here has also been reduced to its bare essentials. I've included only the topics that I feel are relevant to the majority of us as we go about our daily business, and these I've taken great pains to present as simply as possible without rendering them meaningless. Formulas, equations, and numbers of any kind have been virtually eliminated. Illustrative examples drawn from everyday experiences and household objects, on the other hand, have been extensively employed. To help you more fully appreciate the discovery or application of certain scientific principles I've included a number of "thought experiments." With a bit of imagination you'll be able to relive the past, more fully experience the present, and even look into the future.

I've also done what I could to eliminate jargon. That's not to say there are no technical terms here. Quite to the contrary, there are quite a few. That's because science is an exact discipline and requires words that have exact meanings. But I've introduced only those terms I considered necessary for clarity, and these I've taken great pains to define thoroughly in plain English.

Perhaps most importantly, I've presented this shortcut, streamlined science in a historical manner. There are several reasons for this. By sharing the discovery process I feel you get a better idea of how we've come to know what we do about the physical world and how the scientific method works. By following the triumphs and defeats of those who explore the physical world in earnest you also realize that science isn't about things so much as it is about people. And finally, I think our continual struggle to solve the world's mysteries is a riveting drama and I'd like to share it with you. Science, after all, is a story, not a subject.

So this is science intended to enlighten you, not to frustrate you. This is science that is basic enough to understand yet useful enough to appreciate. This is science designed to be entertaining or, at the very least, interesting. This is science that will help you to better notice, marvel at, and enjoy the universe.

Richard Paul
April 1993

A HANDBOOK TO THE
UNIVERSE

WHAT'S THE MATTER?

GLUE RESIDUE

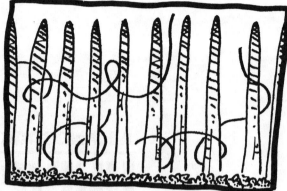

COMB CRUD

DRYER LINT

SHOE FLOP

POCKET SCHMUTZ

NOTHING, NOTHING AT ALL

MCF

I MATTER

"By convention there is color, by convention sweetness, by convention bitterness, but in reality there are atoms and space."
—Democritus, c.460–c.370 B.C.

I.1 A MATERIAL WORLD

Probably the most obvious feature of planet Earth is its amazing amount and assortment of stuff. Stuff is everywhere—all kinds of it. Hot, cold, wet, dry, hard, soft, big, small, quick, dead, new, used, foreign, and domestic. Stuff in just about every size, shape, color, texture, and price range you can imagine. Everything, it seems, is made out of stuff of one kind or another. In all its quantity, variety, and ubiquity stuff is a source of genuine wonder. Where did it all come from? Why is it here? What's it made out of? What in the world *is* all this stuff, anyway?

Well, in the first place, among those who deal with it seriously, it's referred to as matter, not stuff. And matter is usually defined as anything that has mass and occupies space. (See Unit I.13.) There's a lot more to it than that, of course. Matter, it turns out, is an intricate, enigmatic, and ultimately mysterious phenomenon. But even as its innermost secrets recede always just beyond our grasp, we continue to amass knowledge about matter that's theoretically impressive and, perhaps more important, extremely useful. It's this knowledge that has enabled us to grow seedless watermelons, turn peanut butter into diamonds, formulate low-calorie sweeteners, and manufacture both nonstick frying pans and nonskid floors. Our mastery of matter hasn't come easily, though. Or quickly.

For thousands upon thousands of years we were helpless, ignorant captives of a material world that was harsh and unyielding, if not downright hostile. Our existence was little more than a struggle to survive the elements. Through the slow, painful, awkward process of trial, error, and accident, however, we began to learn a few tricks. A tree branch made a good club and certain stones could be fashioned into spear points and arrowheads. Clay could be fired and turned into pottery, plant fibers could be spun and woven into cloth, and special magical rocks, when sufficiently heated, yielded shiny metals that could be hammered into tools and utensils. Slowly, ever so slowly, survival became a bit easier, existence a bit more comfortable and life more pleasant.

This gradual accumulation of manipulative skills, although persistent, was haphazard and extremely tedious. Random, isolated problems were solved, but no organized progress was made. Matter became increasingly useful, but not a bit more understood. Humankind remained a captive; not quite as helpless, perhaps, but just as ignorant.

Then, about 700 B.C. or so, along came the Greeks who, as far as we know, were the first to consider matter in any sort of abstract or systematic manner. This is hardly surprising, for the Greeks had something of a mania for asking and then trying to answer as many difficult questions as they could think of. Questions like: What is truth? What is beauty? What is good? What is evil? And, inevitably, what is stuff?

It was Thales of Miletus (c.624–c.546 B.C.) who evidently gave the earliest serious attention to matter. After much thought he concluded that everything in the world was made exclusively out of water in one form or another. A bit simplistic, to be sure, but a starting point nonetheless. The most important thing Thales did was introduce the idea of an element—a simple, basic substance out of which other, more complex substances were made. Anaximenes (6th century B.C.) was quick to grasp the idea and proposed air as the universal ingredient. Heraclitus (c.540–c.480 B.C.) favored a world made out of fire.

A new twist was added by Empedocles (c.490–430 B.C.). Perhaps in an attempt to please everyone, he introduced the concept of multiple elements. Thales could have his water, Anaximenes his air, and Heraclitus his fire. And, perhaps to make sure his efforts were remembered, Empedocles added a fourth element of his own as well: earth. According to these earliest explorers of the material world, there were really only four basic kinds of stuff when you got right down to it. Everything was made out of water, air, fire, or earth in various combinations.

1.2 A PRIMITIVE THEORY

Having pretty well agreed upon what matter was made up of, the Greeks needed only to determine how it was put together in order to round out their primitive theory. Democritus (c.460–c.370 B.C.) had what he thought was a pretty clever idea regarding this issue. Matter, he said, was an assembly of minute particles so small that they could be no smaller. He called his invisibly tiny particles atoms, which in his day meant "indivisible." Matter, then, rather than being whole and solid throughout, as it appeared, was really made up of discrete little crumbs with gaps among them. Each crumb was an atom, the basic unit in which matter existed. There could be no amount of matter smaller than an atom's worth and no amount between one atom's worth and two. Larger quantities existed only as assemblages of these basic pieces. Just as the population of Athens consisted of numerous individual and indivisible citizens, so the matter in a chunk of stuff consisted of numerous individual and indivisible atoms.

To accommodate the four elements of Empedocles, Democritus proposed four kinds of atoms, each with its own distinctive shape. As far as he was concerned everything was made up of, and could be broken down into, water atoms, air atoms, fire atoms, and earth atoms. It was the combination of these four varieties of atoms in varying ratios and patterns that accounted for all the different types of matter. Each substance had its own special mix and arrangement of atoms.

Besides being indivisible, Democritus thought, these atomic building blocks were also indestructible and immutable. A fire atom would always be a fire atom, no more, no less, even though it might unite with other atoms to form other, more complex substances. It might, for example, start out as part of

■■■■■■■■■■■■■■■■■■■■■■■■■■■■■■■■■■■■■■

Aristotle

Born in Asia Minor, Aristotle went to Athens at the age of 17 to study under Plato. This he did for 20 years, until Plato's death. He was then called upon to tutor the young Alexander, the son of King Phillip of Macedonia. It was Alexander who, after conquering the world and becoming "The Great," arranged for Aristotle to have his own school, the Lyceum.

It was at the Lyceum that Aristotle indulged his insatiable curiosity about everything from drama and athletics to psychology and biology. During his 13 years there he authored writings that were of encyclopedic proportions and scope. They were destined to have an unequalled impact on the evolution of Western thought.

Aristotle enjoyed his greatest popularity during the depths of the Dark Ages, when he was considered the foremost authority on just about everything and often referred to simply as "The Philosopher." His ideas about the operation of the physical world are now ridiculously outdated, but he is still regarded as something of an authority in the areas of ethics and politics.

■■■■■■■■■■■■■■■■■■■■■■■■■■■■■■■■■■■■■■

a tree branch, become firewood, escape from a kitchen fire as a puff of smoke, be washed back to earth in a thunderstorm, and then become part of another tree. During the whole cycle it would retain its identity as a fire atom. It would never become a water atom, an air atom, or anything else.

Although acclaimed as something of a prophet by modern scientists, Democritus was ridiculed as something a crackpot by his contemporaries. The most troublesome of his detractors was a fellow named Aristotle (384–322 B.C.), one of the most highly respected of all Greek thinkers, especially in the area of natural philosophy—the study of the physical world. At his own personal think tank in Athens this intellectual overachiever accumulated and cataloged all the knowledge he could lay his hands on and added to it a good deal of original material. By the time he was through there was little for anyone else to do but read the results. The subject of matter received thorough coverage.

Aristotle had little use for Democritus's atoms—found them a bit contrived, actually. He agreed with Democritus that complex substances were made out of simple ones, but that's where they parted company. Both, for instance, might have agreed that figs were a mixture of earth and water with maybe just a dash of fire thrown in. In Democritus's mind, though, a fig could only be chopped up so far until it lost its identity as a fig and became once again the bits of earth, water, and fire of which it was made in the first place. Aristotle, meanwhile, thought that once earth, water, and fire became a fig they completely lost their original identities. For him there was no limit to how finely a fig could be fragmented, and the parts of a fig, no matter how small, would retain their figness.

To Aristotle the concept of indivisibly small particles implied a grainy, disjointed reality and there was, he felt, no reason to invent such an awkward notion. Everything within the realm of everyday experience suggested a smooth and uniform world, not one existing in bits and pieces. Time, space, and motion all apparently flowed without interruption, and matter should too. Why bother to break it up into little indestructible chunks and call them atoms?

The four elements of Empedocles, on the other hand, were heartily endorsed by Aristotle. Four was a nice, even, geometrically pleasing number. It was so nice, in fact, that Aristotle decided to give to the four elements four fundamental properties: heat, cold, humidity, and dryness. Each element was said to be naturally imbued with a pair of these properties. Water was cold and wet, air hot and wet, fire hot and dry, and earth cold and dry. All other properties were derived from these four starting points just as all other substances were derived from the original four elements, each according to its own particular recipe. To complete his explanation of matter, Aristotle also assigned to every object various amounts of either gravity or levity. Some objects contained mostly levity, others mostly gravity. It was the levity in flames and smoke that sent them skyward. It was the gravity content of rocks that determined their weight and the speed at which they fell. Heavy rocks, containing more gravity than light ones, fell faster.

Starting with nothing more than these four elements, four properties, and gravity and levity, Aristotle was prepared to address just about any issue concerning matter that came his way. Being a quick thinker and a good talker, his answers were invariably elaborate and often quite eloquent. But, having nothing more to base his arguments on than a set of assumptions totally lacking any factual basis, his answers were also usually more than a little bit vague and almost always altogether worthless as far as any practical applications were concerned.

Take combustion, for instance. Combustion, the transformation of wood into smoke, heat, light, and ash, was widely acknowledged to be one of the most spectacular events that took place in the material world. Any theory of matter worth debating had to be able to explain this fantastic phenomenon. The Aristotelian version was that the elements embodied in the wood—some sort of concoction of water and earth—underwent a rather violent transformation into the element fire. The coldness and wetness of the wood turned into the hotness and dryness of the fire while the gravity of the fuel became the levity of the flames. Fine and dandy, but how did all this happen? And why? Nothing Aristotle had to offer was of any use whatsoever to anyone seeking a more efficient fuel, a hotter flame, or a better fire extinguisher.

But let's be fair. The Greeks were philosophers, not scientists. Asking questions, not answering them, was their forte. And for this we must give them credit. They were the first ones to ask the questions we're still trying to definitively answer. What is matter? What's it made out of? And how is it constructed?

I.3 A SUFFOCATING STATUS QUO

Eventually, for reasons that will forever be debated, the Greek Empire was absorbed and replaced by that of the Romans, whose attitudes about matter tended to be of a pragmatic rather than contemplative sort. They were concerned about what kind of stuff was best for paving the Appian Way or how much of it was needed to feed the Second Legion, but didn't fuss much over its hidden nature. They accordingly performed spectacular feats with stone and sword, but scarcely even tried to advance the ideas of Aristotle and his associates. When their empire in turn crumbled, sometime around 400 A.D., the only ideas about matter the Romans had to pass along were those they had inherited from the Greeks. The physical world, just as it had been for the previous 700 years or so, was still assumed to be constructed entirely out of four basic ingredients: water, air, fire, and earth.

And so it was destined to remain a good while longer. As history would have it, the Romans were survived by no heirs apparent. Western civilization, left without a custodian, rapidly began to deteriorate. Anarchy replaced empire, Latin gave way to a babble of foreign tongues, and the bright glory of Rome faded into the dimness of the appropriately named Dark Ages. Just as it had been way back before the rise of Greece, day-to-day survival once again became humanity's foremost concern, and that meant maiming and plundering your neighbors before they maimed and plundered you.

Under these conditions the issue of whether matter was creamy or chunky was of relatively minor importance and rarely debated. A few theologians were just about the only ones interested in scholastic matters, and their goal was simply to maintain what knowledge there was, not advance it. While marauding tribes warred over one another's barley, horses, and women, these monks, sequestered in their monasteries, spent their days reassembling the scraps of wisdom, mostly of Greek origin, that had been littered across Europe, Asia, and northern Africa after the fall of Rome. This by now ancient, incomplete, and adulterated body of thought they selectively translated, interpreted, and passed along in a manner consistent with Church policy.

Because they were considered fairly benign, Aristotelian ideas about matter survived pretty much intact and indeed, once sanctioned by the Church, assumed quasi-ecclesiastic status. Like the triune God and the Ten Commandments, the four elements became a higher truth. Any original thought on the subject was strongly discouraged; progress virtually prohibited. Years, decades, and centuries passed and still the stuff of the physical world was assumed to consist of four basic elements, to have four primary properties, and to be possessed of either gravity or levity.

This long-standing allegiance to the repressive Church, in conjunction with total reliance on the woefully mistaken Aristotle, produced an atmosphere of unprecedented intellectual frustration and fostered the rather bizarre practice known as alchemy. A strange mix of embryonic science, religion, philosophy, and magic, alchemy was behind the obsessive pursuit of something called the Philosophers' Stone. This stone, if ever found, was supposed to have been able to transform common metals like lead into precious ones like gold as well as serve as the "elixir of life," the cure for all bodily ills including old age.

Motivated more by greed than curiosity, the alchemists poked, powdered, boiled, and baked just about everything possible in their search for the Philosophers' Stone and the riches and glory that would surely accompany it. They made some useful discoveries and refined the practices of pharmacology, glassmaking, casting, smelting, dyeing, and tanning in the process but, because they were handcuffed by the Aristotelian-Christian version of matter, their gains in those areas were rigidly circumscribed. Their theoretical accomplishments, which would have required shedding their manacles, were nil. During a thousand years of lab work they succeeded in little more than accelerating the ancient process of trial and error. Matter remained as perplexing as ever. Combustion was still a complete mystery.

This suffocating status quo lingered on until the arrival of that wonderful, inexplicable period called the Renaissance. Gradually, then more rapidly, new ideas began to sprout, blossom and, with the help of Johannes Gutenberg's printing press, cross-pollinate. Christopher Columbus found a brand-new world, Martin Luther found a new way to heaven, and a freethinking Italian named Galileo Galilei (1564–1642) found out that Aristotle was just as human as anyone else.

For almost 2,000 years it had been submissively believed that heavier objects, possessing more gravity, fell faster than light ones. In a bold move, Galileo decided to test this age-old axiom, something no one else—certainly not Aristotle—had ever bothered to do. The experiment he performed was simple. As legend would have it, Galileo lugged a relatively heavy object and a second much lighter object to a high balcony and then dropped them side by side. The results he achieved were dramatic. The heavy object, the light object, and Aristotle's credibility came crashing down with a simultaneous thud. (See Unit II.5.) When the dust cleared, an age of dogma had given way to an age of discovery.

I.4 DOCUMENTED FACTS

The repercussions of Galileo's impudence were immeasurable. Aristotle had clearly been wrong about gravity and levity. Maybe he had been mistaken about the four elements as well. Maybe there were three elements. Or five. Or five hundred. Maybe there really were atoms. Maybe it was possible to find out.

■■■■■■■■■■■■■■■■■■■■■■■■■■■■■■■■■■■■

A Legendary Experiment

Galileo's dramatic disproof of Aristotle, perhaps the most famous scientific experiment of all time, is often portrayed as having taken place at the Leaning Tower of Pisa before a rapt audience. There exists not a shred of evidence to support this romantic notion. Then again, it could have happened.

Galielo was born in Pisa the son of a musician, studied medicine there as a young man, and later taught there. It was during his tenure as a professor of mathematics at the University of Pisa, sometime around 1590, that he investigated the behavior of falling bodies. Galileo is also known to have enjoyed being in the limelight.

Regardless of how, where, or for whom he conducted his experiment, it's widely acknowledged that it was Galileo who was chiefly responsible for discrediting Aristotle's logico-verbal method of exploring the physical world and replacing it with a more scientific one. "The book of nature," he declared, "is written in the language of mathematics." Whether it's true or not, the Leaning Tower legend is entirely appropriate.

■■■■■■■■■■■■■■■■■■■■■■■■■■■■■■■■■■■■

Beginning with Galileo, eyewitnesses replaced ancient texts as authorities about how the physical world worked. Experimentation replaced argumentation as the preferred means of investigating it, and those doing the investigating were properly called scientists, not philosophers. The first important one, as least as far as matter is concerned, was Robert Boyle (1627–1691). It was Boyle who first described the behavior of matter in terms of documented facts.

Boyle, the seventh son of the earl of Cork, published his work as *The Sceptical Chymist* in 1661. In this quaint but thoughtful treatise he flatly rejected the four elements invented by Empedocles and authorized by Aristotle. He embraced the idea of elements, but for him they weren't water, air, fire, and earth, but an as yet undetermined number of "certain Primitive and Simple or perfectly unmingled bodies." And, just to make sure the break with Aristotle was complete, he revived Democritus's atoms, calling them by his own name: corpuscles. These, he said, were "little particles, of several sizes and shapes, variously moved."

Boyle's elements were still just vague ideas, and his little corpuscles totally imaginary. He was guessing, just like Empedocles and Democritus had been about their elements and atoms. But there was a big difference. Boyle based his belief in elements upon physical evidence, not a speculative hunch. He had extracted basic, underlying substances from obviously more complex ones. He

• •

The Scientific Method

The first truly scientific breakthrough regarding matter is generally attributed to Robert Boyle, born in England a few years after the departure of the Pilgrims for the New World. Although, like many learned men of his day, he was a theologian, Boyle also had a keen interest in worldly affairs. His specialty was the budding discipline of chemistry, which he studied in his personal laboratory outfitted with thermometers, scales, barometers, calibrated flasks, and an air pump that he had helped develop.

It was while examining the operation of the barometer at pressures far above and below those encountered under normal atmospheric conditions that he noticed something curious. Every time he pushed a stopper down the neck of a flask the air pressure within the flask increased. As he pushed harder on the cork it continued to compress the air inside the flask and the pressure continued to rise. This, he decided, merited further investigation.

His first step was to come up with a hypothesis—a proposed statement about the relationship between air pressure and air volume subject to verification (or rejection) based on facts. He then collected some relevant facts by conducting a number of carefully designed experiments. When his lab findings supported his hypothesis he supposed it to be true and called it a theory. He then published his work and waited for his colleagues to either confirm or refute it.

Boyle's theory, of course, was confirmed. Not only that, but as new gases like oxygen, carbon dioxide, and nitrogen were discovered the theory was found to apply to them as well as to air. Eventually, when the similar molecular behavior of all gases was understood, the reason the theory worked was explained. (See Unit 1.23.) It was then considered a substantiated statement of truth and called Boyle's law.

Boyle's accomplishment was the result of his strict adherence to what is known as the scientific method.

—What do you think determines whether or not something is scientific?

—What makes the scientific method so much more powerful in the search for truth about the physical world than the purely verbal methods employed by the great Greek thinkers?

—Do you think Boyle's original hypothesis was stated in mathematical terms? What about his fact-supported theory?

—What sort of experiments do you suppose Boyle performed?

• •

had obtained a wonderfully flammable powder called phosphorous from nothing more than common urine. Phosphorous, along with other "Primitive and Simple" ingredients, had obviously "mingled" to form the urine in the first place. His belief in particles arose not from rumination, but from observation. Gases were less dense than solids or liquids, he concluded, because their component "little particles" were farther apart.

But despite all the changes in attitude and the improvements in technique, chemistry, as it was now being called, still had a lot more questions than answers. Foremost among the many thorny problems in need of resolution there remained combustion—common burning. Fire was obviously a powerful agent, capable of reducing a block of solid wood into a pile of practically weightless ashes and a wisp of smoke. What happened to all the wood? The question, thousands of years old, just wouldn't go away. In light of the prevailing mood a workable, verifiable answer was needed. Animal spirits, fire gods, and magic were no longer in vogue.

To solve the ancient riddle a German scientist, Georg Ernst Stahl (1660–1734), dreamed up a substance he christened phlogistom. This, he said, was an odorless, tasteless, invisible ingredient present in combustible materials like wood. The act of combustion liberated the captive phlogistoñ. The ashes of a fuel weighed less than the original fuel by the amount of phlogiston lost during combustion. Even though no one could actually detect it, the existence of phlogiston seemed to explain the available facts. Because no one could really prove it wrong, the phlogiston theory became quite popular.

But wood, it turns out, isn't the only thing that burns. Flesh and bone burn, sugar burns, coal and tar burn, and in a hot enough flame metals like mercury and tin burn. And, rather interestingly, the ashes of burned metals weigh more, not less, than the original metals. This fact was pointed out to Stahl. Ever creative, he explained that phlogiston sometimes had a negative weight. On these occasions its absence after combustion resulted in a net weight gain rather than loss.

Most of Stahl's contemporaries, having no better alternative, continued to believe in his farfetched, schizophrenic, phantom phlogiston. Most, but not all.

I.5 | FUNDAMENTAL TYPES OF MATTER

To Antoine-Laurent Lavoisier (1743–1794), phlogiston was just another fancy name for a fire god. The French aristocrat, lawyer, gunpowder manufacturer, tax collector, and chemist made up his mind to find a more plausible, more tangible explanation of combustion.

Lavoisier had a few things going for him in this endeavor. To begin with he had information about, and samples of, two intriguing new "airs": fixed air, discovered by Joseph Black (1728–1799), and dephlogisticated air, discovered

■■■■■■■■■■■■■■■■■■■■■■■■■■■■■■■■■■■■■

Antoine-Laurent Lavoisier

Lavoisier was born privileged in Paris. His aristocratic heritage was the result of his father's purchase of a title of nobility, a common practice among the wealthy. He was sent to the finest schools and studied mathematics, astronomy, chemistry, and botany as well as law.

Lavoisier had an incredible range of interests and activities. He established experimental farms, developed streetlighting plans, helped formulate the metric system, and participated in other affairs as varied as coinage, insurance, and drainage. His association with gunpowder was the result of his being placed in charge of the French arsenal. In his free time he wrote treatises on everything from thunder to divining rods.

Lavoisier collected taxes under a royal license prior to the French Revolution, and after it was accused of cheating the government. He was sentenced to death by a revolutionary tribunal and guillotined. Perhaps as a consolation, he's now immortalized as a hero of the chemical revolution—in which he fought on the winning side.

■■■■■■■■■■■■■■■■■■■■■■■■■■■■■■■■■■■■■

by Joseph Priestley (1733–1804). Both of these had been extracted from ordinary atmospheric air, thereby refuting Aristotle's claim that air was an irreducible element. Lavoisier also had sensitive scales that allowed him to monitor his experiments with unprecedented accuracy.

Lavoisier's first major finding was that combustion, when confined within a sealed flask, resulted in no weight gain or loss of any kind. A closed flask of anything, be it wood, sugar, or mercury, weighed precisely the same before and after its contents were incinerated. The contents of the flask were drastically altered, but nothing, it seemed, was absolutely created or destroyed during the act of combustion.

He then determined that combustion, rather than being a release of phlogiston *into* the atmosphere, was actually the absorption *from* the atmosphere of Priestley's dephlogisticated air, which he mercifully renamed oxygen. The ashes of a metal burned in open air outweighed the original metal by the weight of oxygen taken up from the surrounding atmosphere. Fire turned mercury into mercury oxide, a combination of mercury and oxygen. The weight of the resulting mercury oxide was equal to the weight of the original mercury plus the weight of oxygen captured from the air. In a sealed flask the weight gain of mercury oxide over mercury was exactly offset by the weight of oxygen lost by the trapped air. The oxygen, along with its weight, was simply transferred from the atmosphere to the mercury. No negative weights were involved.

In the case of wood, combustion involved the combination of the oxygen in the air with carbon, an important ingredient of most common fuels. This process formed Black's fixed air, better known now as carbon dioxide. During open combustion it was gaseous carbon dioxide, not phlogiston, that escaped into the atmosphere, carrying away with it the originally solid carbon and its weight from the wood. Only the incombustible impurities remained behind as ash. In a closed flask the weight of the trapped carbon dioxide created exactly compensated for the weight of carbon lost by the wood and the weight of oxygen lost by the air. Again, it was merely a rearrangement of ingredients and weights that took place. Phlogiston, no longer needed for anything, was appropriately abandoned.

Lavoisier, freed of crippling misconceptions, continued to experiment and succeeded in running certain chemical reactions forward and backward. He was able, for example, to burn mercury in air to get mercury oxide and then reduce mercury oxide once again to the mercury and oxygen of which it was formed. Every time Lavoisier succeeded in reversing a reaction like this he recovered his original ingredients, not only in their original conditions, but in their original amounts. Based on this work he formulated the law of the conservation of matter, which states that matter can be endlessly altered but never absolutely created or destroyed. This remains a basic principle of chemistry, despite recent modifications. (See Unit II.29.)

Some of Lavoisier's ingredients defied any further attempts at analysis or reduction. Oxygen could be combined with all sorts of other substances to yield a wide variety of products and could, in turn, be extracted from those products again as oxygen. But, try as he might, Lavoisier could never extract anything else from oxygen nor could he find any combination of ingredients that united to form it. Mercury, lead, phosphorous, sulfur, carbon, and a growing list of other basic ingredients also seemed to be starting points for building up other substances and end points for taking them apart.

Here were the "Primitive and Simple or perfectly unmingled bodies" of Boyle. Here were fundamental types of matter common to many other substances but unique unto themselves. Here were the elements first conceived of by Thales, Empedocles, and Aristotle. But these weren't the elements of some philosopher's imagination; these were the elements of a chemist's lab.

Lavoisier published his findings as the *Traité Elémentaire de Chemie* in 1789. In this chemical textbook he stated the "incontestible axiom, that in all the operations of art and nature, nothing is created; an equal quality and quantity of the elements remain precisely the same, and nothing takes place beyond changes and modifications in the combination of these elements." The book was an instant success and subsequently translated into English, Dutch, Italian, and Spanish. The scientific community read about the 23 elements Lavoisier had identified and quickly took up the search for others.

1.6 A HIDDEN MECHANISM

What the flamboyant Frenchman did for the elements in the waning years of the eighteenth century a retiring English schoolteacher did for atoms in the early years of the nineteenth. John Dalton (1766–1844) attended a small, rural, Quaker school for a total of six years. Then, at the venerable age of 12, he became a teacher and headmaster at the same school. It was later, as a professor of physics and mathematics at New College in Manchester, that he developed his atomic theory of matter.

A son of a weaver, a man who for 57 years never missed his daily walk in the Manchester countryside, a colorless figure by all accounts, Dalton seems an unlikely candidate to make a scientific breakthrough. But it was because of, not in spite of, his plodding style that he had success. While his colleagues were off "element hunting," hoping to identify and maybe even name their very own elements, Dalton was methodically scrutinizing those already well known: oxygen, hydrogen, carbon, and a few others. These he endlessly combined, took apart, and recombined, each time, like Lavoisier, carefully measuring what he started out with (called reactants by chemists) and what he ended up with (called products).

Being of a mathematical mind, he recognized some interesting patterns. Hydrogen, he determined, when burned in the presence of oxygen, always produced water and only water. And always two parts, by volume, of hydrogen combined with one part of oxygen. A mixture of the gases in any other ratio merely produced water and some leftover hydrogen or oxygen. Water, it seemed, was a combination of the elements hydrogen and oxygen and nothing else. More specifically, it was the combination of exactly two parts of hydrogen and one part of oxygen. Somewhat analogously, oxygen and carbon combined in a precise and unfailing ratio of two parts to one to form carbon dioxide. Methane, meanwhile, was formed from one part of carbon and four parts of hydrogen.

Dalton was intrigued by these precise proportions and the fact that they were whole-number multiples of one another. The evidence suggested that elements combined not arbitrarily or en masse, but selectively according to some hidden mechanism. When oxygen and hydrogen united chemically to produce water they did so in specific amounts. A unit of oxygen always combined with two units of hydrogen to form water. A unit of oxygen never formed water with a single unit of hydrogen, three units of hydrogen, a half unit of hydrogen, or any other fractional amount. The units involved were whole and indivisible. The units were the fundamental quantities in which the elements existed and behaved. The units, in the final analysis, were atoms.

Democritus had been right all along—matter was chunky. Just as he had theorized a couple thousand years earlier, elements existed as little particles. It was the linking together of selective numbers of these particles in various ways that produced complex substances from simple ones. Two atoms of hydrogen evidently united with an atom of oxygen to form water and four atoms of hydrogen combined with an atom of carbon to form methane. Burning mercury to form mercury oxide, it followed, involved combining atoms of mercury with atoms of oxygen from the air.

The atom, again as Democritus had predicted, was also apparently indestructible and immutable. When mercury was recovered from mercury oxide it was the very same atoms of mercury that had reacted with oxygen to form the mercury oxide in the first place that were being recaptured. These atoms remained intact even while undergoing drastic chemical changes. The recovered atoms, of course, could be used to make mercury oxide again or any other complex substance that required mercury as a basic ingredient. Atoms of oxygen, mercury, lead, carbon, and however many other elements existed were constantly being put together, taken apart, exchanged, and rearranged as the physical world underwent its assorted changes and processes. Or, as Dalton put it, "All the changes we can produce consist in separating particles that are in a state of . . . combination and joining those that were previously at a distance."

Dalton went on to determine the ratio of elemental ingredients of many other common substances and went so far as to sketch their "compound atoms." The compound atom—what we now call a molecule—of water, for instance, he depicted as consisting of two atoms of hydrogen and one atom of oxygen glued together in a little clump. And, as if that weren't enough, he then came up with a table of atomic weights. Working always with bulk amounts, he discovered that the atoms of each element had their own particular weight. The lightest atoms were those of hydrogen. These he assigned a weight of 1. All other atoms had weights that were multiples of those of hydrogen and were measured accordingly. Carbon atoms, for example, weighed 12 times as much as those of hydrogen and so were assigned a weight of 12, and so on.

By the time Dalton was finished with his momentous work he had pretty well developed the atomic theory of matter that, with a few modifications and elaborations, has endured to the present day. In summary, as he detailed in *A New System of Chemistry,* published in 1808:

—All matter, everything from dust to dichlorodiphenyltrichloroethane, is made up of a select number of basic ingredients called elements, as first suggested by Thales and Empedocles and later substantiated by Lavoisier.

—Each of these elemental ingredients exists as a collection of tiny, immutable, indivisible, and indestructible little particles called atoms, just as Democritus had imagined. (See Units I.26–28 for recent developments regarding the mutation and destruction of the atom.)

—The atoms of any particular element are identical to one another in every way.

—Different elements have different kinds of atoms. In particular, the atoms of different elements differ in weight, and can be identified on that basis.

—Chemical reactions are merely a rearrangement of atoms.

—The formation of complex substances from their elemental ingredients takes place through the formation of "compound atoms" containing a definite number of atoms of each of its elements.

I.7 AN UNDERLYING ORGANIZATION

With the existence of elements well substantiated, the combustion riddle satisfactorily solved, and the atomic theory firmly in place, many nineteenth-century chemists felt there was little left for them to do but hunt the elements to extinction. Hundreds of complex substances were analyzed to determine their underlying ingredients. Ingredients that resisted further analysis were proclaimed elements. By 1870 a total of 63 elemental trophies had been stuffed and mounted on laboratory walls. Amid much international oohing and aahing each was scrutinized, admired, weighed, and named.

A Swedish chemist, Jöns Jacob Berzelius (1779–1848), went so far as to provide each element with a nickname, a one- or two-letter designation that amounted to chemical shorthand. When possible, elements were tagged simply by their initials. Hydrogen was H, boron B, and carbon C. When two or more elements shared an initial a second letter was added for clarity. Helium was He, barium Ba, and cerium Ce. A few old-timers were given symbols derived from their ancient, usually Latin or Greek, names. Gold was Au, from *aurum,* silver Ag from *argentum,* and tin Sn from *stannum.*

As Dalton had predicted, each element was composed of atoms of a distinctive weight. By the weight of its atoms any element could be identified and placed in sequential order, starting with number 1, hydrogen, whose atoms were the lightest known. But despite their unique identities, many elements shared remarkably similar properties. Sodium and potassium, for instance, were both whitish and extremely active metals that burned with a dazzling brilliance. Gold and silver, meanwhile, were both extremely stable, chemically inactive metals that were relatively soft and highly malleable. All sorts of properties, in fact, appeared periodically although irregularly throughout the elemental lineup.

Science had come too far by this point to attribute anything of this sort to mere coincidence. An underlying organization of the elements was presumed and, of course, chemists around the world went to work trying to figure out what it was. They plotted atomic weights on a variety of graphs, analyzed the size of the weight intervals between adjacent elements, and counted the number of elements between those with similar properties. But nothing solved the riddle of peridocity.

The answer was finally supplied by Dmitri Ivanovich Mendeleev (1834–1907), a Russian professor of chemistry at the University of St. Petersburg. Possessing unheard-of quantities of patience, energy, determination, and persistence, he was the perfect man for the job. After years of relentless effort he

• •

The Periodic Chart

Dmitri Ivanovich Mendeleev, the man who established law and order among the previously unruly mob of elements, was well acquainted with adversity. He was born on the fringe of civilization in Tobol'sk, Siberia, the youngest of 17 children. At the age of 15 he traveled by horse and wagon to distant Moscow seeking admission to the university there, only to be rejected as being "academically and intellectually unfit." He overcame these early obstacles with hard work and made sense out of the periodic recurrence of properties among the elements the same way.

His first step was to assemble every known fact about every known element. Then, to make these data easier to manipulate, he placed them on a series of cards, one for each element. These cards he endlessly shuffled, looking always for a pattern of some sort. He arranged his cards in spirals, circles, and pyramids, firmly believing that a design would eventually emerge.

After countless failures and immeasurable frustration, it more or less did. It took the form of a strange-looking chart, with both columns and rows of uneven sizes. With the exception of placing the heavier tellurium ahead of lighter iodine, it ranked the elements in order by weight. By leaving a number of empty spots along the way it also managed to place similar elements in vertical groups or "families." In defense of its apparent shortcomings Mendeleev boldly postulated the existence of as yet unknown elements to fill its gaps, and claimed that the accepted weight of either iodine or tellurium must be wrong.

His confidence turned out to be well founded. Gallium, scandium, and germanium were soon discovered with the weights and properties that Mendeleev predicted. Later, when the structure of the atom was deciphered, even his flip-flopping of iodine and tellurium was justified. (See Unit 1.27.)

A modern version of Mendeleev's creation is essential to anyone working with the elements. (See Figure 1.1.)

—How many of the elements are familiar to you? Do they seem to occupy only certain areas of the chart?

—Can you guess what other elements are metals, based on those you already know?

—Carbon is the main ingredient in most plastics. What other element do you think might be used for this purpose?

—Sodium (Na) and chlorine (Cl) combine to make table salt. What elements do you think combine to make similar substances?

• •

PERIODIC TABLE OF THE ELEMENTS

GROUP																	
IA	IIA	IIIB	IVB	VB	VIB	VIIB	VIII			IB	IIB	IIIA	IVA	VA	VIA	VIIA	0
1 H Hydrogen																	2 He Helium
3 Li Lithium	4 Be Beryllium											5 B Boron	6 C Carbon	7 N Nitrogen	8 O Oxygen	9 F Fluorine	10 Ne Neon
11 Na Sodium	12 Mg Magnesium											13 Al Aluminum	14 Si Silicon	15 P Phosphorus	16 S Sulfur	17 Cl Chlorine	18 Ar Argon
19 K Potassium	20 Ca Calcium	21 Sc Scandium	22 Ti Titanium	23 V Vanadium	24 Cr Chromium	25 Mn Manganese	26 Fe Iron	27 Co Cobalt	28 Ni Nickel	29 Cu Copper	30 Zn Zinc	31 Ga Gallium	32 Ge Germanium	33 As Arsenic	34 Se Selenium	35 Br Bromine	36 Kr Krypton
37 Rb Rubidium	38 Sr Strontium	39 Y Yttrium	40 Zr Zirconium	41 Nb Niobium	42 Mo Molybdenum	43 Tc Technetium	44 Ru Ruthenium	45 Rh Rhodium	46 Pd Palladium	47 Ag Silver	48 Cd Cadmium	49 In Indium	50 Sn Tin	51 Sb Antimony	52 Te Tellurium	53 I Iodine	54 Xe Xenon
55 Cs Cesium	56 Ba Barium	57–71 Lanthanides (see below)	72 Hf Hafnium	73 Ta Tantalum	74 W Tungsten (Wolfram)	75 Re Rhenium	76 Os Osmium	77 Ir Iridium	78 Pt Platinum	79 Au Gold	80 Hg Mercury	81 Tl Thallium	82 Pb Lead	83 Bi Bismuth	84 Po Polonium	85 At Astatine	86 Rn Radon
87 Fr Francium	88 Ra Radium	89–103 Actinides (see below)															

TRANSITION ELEMENTS

Lanthanide Series	57 La Lanthanum	58 Ce Cerium	59 Pr Praseodymium	60 Nd Neodymium	61 Pm Promethium	62 Sm Samarium	63 Eu Europium	64 Gd Gadolinium	65 Tb Terbium	66 Dy Dysprosium	67 Ho Holmium	68 Er Erbium	69 Tm Thulium	70 Yb Ytterbium	71 Lu Lutetium
Actinide Series	89 Ac Actinium	90 Th Thorium	91 Pa Protactinium	92 U Uranium	93 Np Neptunium	94 Pu Plutonium	95 Am Americium	96 Cm Curium	97 Bk Berkelium	98 Cf Californium	99 Es Einsteinium	100 Fm Fermium	101 Md Mendelevium	102 No Nobelium	103* Lw Lawrencium

Artificial or transuranic elements

1
H
Hydrogen

↑ Atomic number
↑ Element symbol
↑ Element name

Figure I.1   A modern periodic chart of the elements

succeeded in organizing the elements in a manner that presented them in sequence by their weights and in groups by their properties. The result wasn't pretty, but it worked. It was the very first periodic chart of the elements, the modern version of which maintains a stern, steady gaze over chemistry students around the world.

Although it might seem rather straightforward or even obvious to us in hindsight, the organization of elements by weights and properties simultaneously was really an exceedingly brilliant accomplishment. Consider the fact that while the elements proceed regularly from one to the next by weight, the properties Mendeleev was interested in occur sometimes at intervals of every second element, sometimes every eighth element, and sometimes every eighteenth or thirty-second element. What's even more impressive is that only about two out of every three elements were known in Mendeleev's day. Developing the chart amounted to winning a complex game of solitaire while playing with no knowledge of the rules and an incomplete deck.

Equipped with his precious invention, Mendeleev boldly and successfully predicted the existence of as yet undiscovered elements and left gaps in his chart to accommodate them. He foretold not only their existence but their weights and their properties as well, with uncanny accuracy. He could even forecast which elements would react with which and, to an extent, how. As if by magic he could name and describe substances no one had ever even imagined, much less seen or heard of. But it wasn't magic, it was the beginning of humankind's mastery of matter.

I.8 | A FULL DECK

The gaps in Mendeleev's periodic chart have now all been filled. We've located and identified all the elements. We have a full deck. How can we be sure? As easy as one, two, three.

The atoms of any particular element differ from the atoms of all other elements by weight, as Dalton discovered. A better way of identifying an atom now, however, is by its atomic number, which corresponds closely to its weight but is actually a function of its complexity. (See Unit I.27.) According to the dictates of atomic architecture, an atom with a number less than 1 is an absurdity and an atom with a number higher than 92 is so complex that it quickly disintegrates into a simpler, more stable structure. There are, then, just 92 atomic numbers available. There are a maximum of 92 types of natural atoms representing 92 elements. (Unnatural elements with numbers higher than 92 are discussed in Unit I.28.) We've found them all, from number 1, hydrogen, on up through number 92, uranium.

Ninety-two different types of atoms account for the enormous diversity of matter we encounter and manufacture in much the same way that just 26 different letters account for the thousands and thousands of words in the English language. They also evidently account for the matter we've yet to encounter. Besides having its atomic weight and atomic number as distinctive characteristics, you see, an element can also be identified by the color of light it emits. Each element, when heated, gives off a characteristic "fingerprint" of colored light that sets it apart from all other elements. The analysis of light from the other planets, the sun, and the far-flung stars and galaxies of space by a process known as spectroscopy reveals what they're made of as surely as would an on-site visit. All are composed of the same 92 basic types of stuff we have here on planet Earth.

The universe as a whole consists mainly of elements number 1 and 2: hydrogen and helium. Here on earth oxygen, number 8, is the most abundant element, comprising just about one-half the mass of the planet. Together with silicon, aluminum, iron, calcium, sodium, potassium, and magnesium it makes up about 97 percent of all earthly matter. The remaining 3 percent includes such nearly unpronounceable substances as protactinium, ytterbium, dysprosium, and praseodymium, to name a few. Some, like terbium, have no known practical uses and would not be much missed in their absence. A few, like, gadolinium, don't exist in quantities large enough to be of any consequence. The world's supply of astatine, element number 85, could be comfortably stored in a thimble.

Other elements, though, are household words. Little bottles of calcium and iron reside in many kitchens where they're used as dietary supplements. Iodine is a standby in many medicine cabinets where it's used as a disinfectant. It's also added to table salt to prevent goiters. Mercury can be found in many thermometers, and lead is what makes car batteries so heavy. Neon glows in the decorative signs of many bars and pizza parlors. Flourine is added to toothpaste and drinking water to ward off tooth decay. Pennies and electrical wiring are made out of copper. Nickels, not coincidentally, contain nickel.

Other elements, just as indispensable, are less well known. Argon, being incombustible, is used to fill light bulbs. Light bulb filaments are made out of tungsten, which is extremely tough. Uranium is used to fuel submarines and atomic power plants. (See Unit II.30.) Cesium allows atomic clocks to keep virtually flawless time. (See Unit IV.12.) In hospitals, radium is used to treat cancerous tumors and barium is an important diagnostic tool.

Only two elements, mercury and bromine, exist as liquids at room temperature. About a dozen, including xenon, helium, and nitrogen, are gases. The rest are solids, most of which are classified as metals. Gold and copper are the only two yellow, or colored, metals. The others, like tin, zinc, and platinum, are white or silvery metals, as is silver itself.

A few elements exist in more than one form. Oxygen, for example, exists primarily as the common gas that makes up some 20 percent of our atmosphere. It also exists in much smaller amounts, however, as ozone, also a gas. Similarly, carbon can take on the guise of either graphite or diamond. These multiple personalities of a single element are called allotropes. Allotropes result when the individual atoms of an element link up in different geometric patterns. Each pattern constitutes an allotrope. (See Figure I.2.)

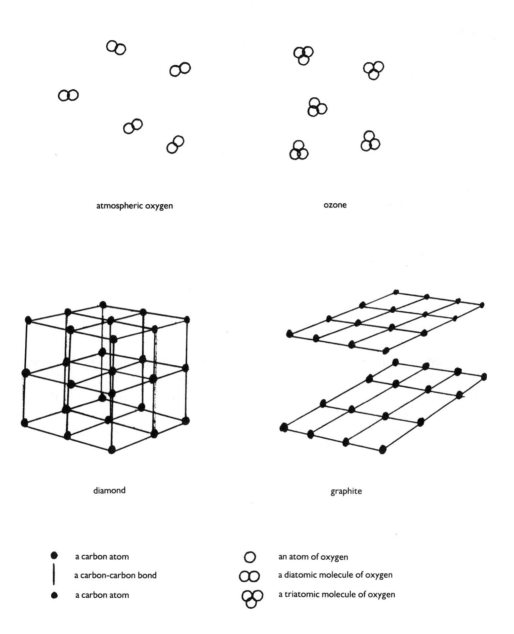

Figure 1.2   Allotropes

Atmospheric oxygen, that benevolent supporter of life, exists as diatomic molecules—molecules consisting of two oxygen atoms glued together like Siamese twins. Ozone, a component of air pollution that is considered a health hazard, exists as triatomic Siamese triplets. Because oxygen atoms can't stand isolation, they're never found in a single, unattached state. Neither are they found in clumps of four or more.

Graphite, commonly encountered as pencil lead, results when carbon atoms line up in two-dimensional grids, like the lines on a sheet of graph paper. There are connections among the atoms on each sheet, but none between atoms of separate sheets. A diamond, in contrast, is a three-dimensional carbon-atom grid. Connections among atoms go every which way, like the girders in the Eiffel Tower. The smeary black stuff in a pencil and the dazzling jewel in a pendant are made out of the same raw material. The only difference between them is the arrangement of their identical atoms. (See Unit I.16.)

I.9 | COMPLEX SUBSTANCES

Like the individual playing cards in a deck or the individual letters in the alphabet, the elements, even with their assorted allotropes, have pretty limited capabilities when considered one at a time. It's when letters combine to form words and sentences that language takes on richness and meaning, and it's when elements combine to form complex substances that matter becomes wonderfully diverse and potent.

And elements just love to combine. So much so, in fact, that they're seldom encountered in isolation. A pure-carbon diamond, an occasional chunk of sulphur, or a rare nugget of gold or copper represent just about the only examples of elements existing naturally in an unadulterated state. This disdain for solitude is one reason why the elements avoided discovery for so many years. Even the scrutiny of the alchemists succeeded in isolating only a dozen or so elements. The rest had to be ferreted out by Lavoisier and his followers with substantial knowledge and specialized equipment.

When elements get together they do so on an atom-to-atom basis, as first described by Dalton. It's individual atoms that meet and, if the chemistry is right, unite or, more properly, bond to form little clumps of atoms known as molecules. Not all elements are compatible, however, and not just any group of atoms will bond as a molecule. Some atoms, like those of hydrogen, are extremely sociable and combine eagerly and rather indiscriminately with all sorts of other atoms, forming a host of complex substances in the process. Then there are others, like krypton atoms, that are fanatically standoffish or, in chemical terminology, inert. In between lies a whole spectrum of different activity levels. But regardless of how gregarious or shy they might be, almost all atoms prefer the company of some partners to others, and are constantly breaking off old relationships to form new ones.

To those who understand it, Mendeleev's periodic chart of the elements is indispensable in understanding this atomic mate-swapping. From an element's position in the chart it's possible to tell which other elements it will combine with and the strength of the resulting chemical union. By comparing the strength of one chemical union to that of another it's also possible to predict chemical reactions. Weaker associations will be abandoned when opportunities to form stronger ones come along. Hydrogen and carbon atoms reside comfortably knitted together in gasoline but, when provided with a motivating temperature and available oxygen, violently reject one another's company. Each prefers a union with oxygen in which they form water and carbon dioxide respectively.

The substances atoms form when they unite chemically to form molecules are properly called compounds. Molecules are the basic particles of compounds in much the same way that atoms are the basic particles of elements. A molecule is the smallest part of a compound that retains the identity and properties of that compound. Just as elements can be chopped into pieces no smaller than a single atom, compounds can be chopped into pieces no smaller than a single molecule. Molecules can be further reduced chemically into their component atoms, of course, but when this happens the compound is destroyed and a whole new substance, or substances, takes its place.

Like the atoms of an element, every molecule of a given compound is identical. A molecule of acetylsalicylic acid, also known as aspirin, in a tablet of Anacin is no different from a molecule of acetylsalicylic acid in a tablet of Bayer, Good 'n' Cheap, or generic aspirin. A molecule of ascorbic acid, also called vitamin C, is only a molecule of ascorbic acid regardless of whether it comes from an orange tree grown in California with organic fertilizers or a smoke-belching factory in New Jersey. (While the molecules of the active ingredient are interchangeable in these instances, molecules of other substances may also be present and affect the purity and/or quality of the final product.)

There's virtually no limit to how large or elaborate a molecule can be. A hemoglobin molecule, for instance, is an intricate complex of nearly ten thousand atoms. Remove, add, or change the identity or position of a single atom and you no longer have hemoglobin; you no longer have a molecule that will make blood red or carry nourishing oxygen to the cells of the body.

The formation of more than one complex substance from the same few elements is fairly common. Under certain conditions hydrogen and oxygen combine in a one-to-one ratio to form hydrogen peroxide rather than in the more familiar two-to-one ratio that forms water. Furthermore, the same elements can combine in the identical proportions to form more than one substance. Six carbon, twelve hydrogen, and six oxygen atoms bond in two different ways to produce the molecules of two distinct sugars: fructose and glucose.

Because it's an entirely new substance, the characteristics of a compound generally bear no resemblance whatsoever to those of its component elements. Sodium is a poisonous metal; chlorine is a poisonous gas. Combine them and

■■■■■■■■■■■■■■■■■■■■■■■■■■■■■■■■■

Isomers

When an identical selection of atoms can be assembled in different configurations to form a variety of molecules, the resulting substances are called isomers. As molecules increase in complexity the number of potential isomers increases dramatically. Large assemblies of atoms are theoretically capable of existing as thousands of isomers. In reality, though, the number rarely exceeds one hundred.

Isomers generally have fairly similar chemical and physical properties. The isomers fructose and glucose, for example, are both simple sugars whose molecules contain six atoms of carbon, twelve of hydrogen, and six of oxygen. Both have the chemical formula $C_6H_{12}O_6$, both are found in fruits and honey, and both are sweet. The differences between them are subtle but of great importance to diabetics, who can tolerate fructose to a far greater extent than glucose.

Octane, C_8H_{18}, exists in 18 isomeric forms, normal or n-octane being the most common. All are flammable, but only the variant iso-octane is used as a standard for measuring the quality of gasoline.

■■■■■■■■■■■■■■■■■■■■■■■■■■■■■■■■■

the result is sodium chloride, better known as table salt, an essential part of a healthy diet. Hydrogen is a dangerously flammable gas; oxygen is a gas necessary for most common types of combustion. When they unite chemically they produce water, an incombustible liquid that is the world's favorite fire extinguisher.

I.10 | A JUMBLE OF PARTICLES

Technically speaking, there are only two categories of true chemical substances: the simple elements and the complex compounds they form when they combine chemically. The elements consist of collections of identical atoms, each with the same atomic number. The compounds consist of collections of identical molecules, each composed of certain atoms in a certain formation. Each element has its own symbol—hydrogen is H, oxygen is O. Each compound has its own formula—water is H_2O. Each substance, whether it be an element or a compound, has its own set of properties that can be used to positively identify it. No two substances are alike. Each is a singular type of matter, a special kind of stuff.

In the course of our ordinary experiences with matter, of course, we don't deal with individual atoms or molecules, but with great milling throngs of them. These throngs are virtually never homogeneous—they almost never

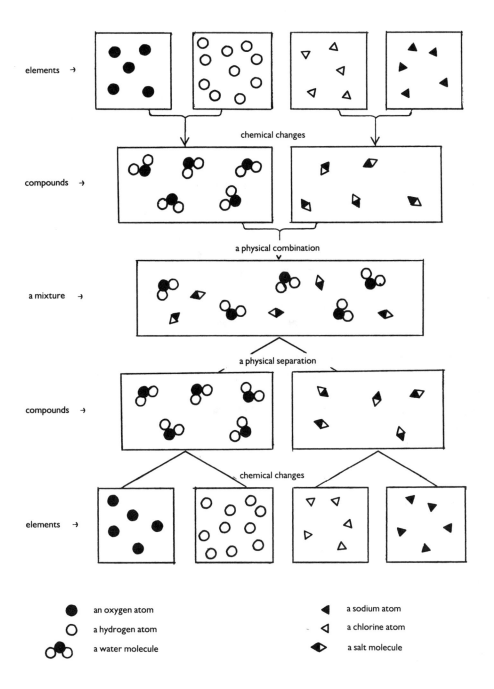

elements →

chemical changes

compounds →

a physical combination

a mixture →

a physical separation

compounds →

chemical changes

elements →

● an oxygen atom

○ a hydrogen atom

a water molecule

◀ a sodium atom

◁ a chlorine atom

◆ a salt molecule

Figure I.3   Chemical versus physical changes

consist of only a single variety of particle. The stuff of everyday life isn't homogeneous, but heterogeneous. It's a jumble of particles, a hodgepodge of substances. In the language of science these hodgepodges are called mixtures.

A mixture, then, consists of bits and pieces of a variety of substances. It's a collection of unorganized atoms and molecules. (This is a good time to point out that an atom is really just a special, simplest possible, type of molecule. From now on, unless indicated otherwise, the word *molecule* will include unattached atoms as well as chemically bonded clumps of them.) While each of its assorted pieces is a substance of some kind or another, the mixture itself isn't a true substance. It has no special formula, no symbol, no atom or molecule that is uniquely its own. A mixture has no set of properties that identifies it or distinguishes it, but instead displays an amalgam of properties that it gets from its component ingredients.

Mixtures are the result of physical, rather than chemical, events. Consider the differences between baking a loaf of bread and tossing a salad. When flour, water, and yeast are put in a hot oven they react chemically to form bread. Their properties as flour, water, and yeast are lost during the process and a totally new substance with its own taste, smell, and texture is created. But when carrots, lettuce, and tomatoes are chopped up and physically mixed in a bowl they remain carrots, lettuce, and tomatoes and are still identifiable as such. The salad isn't really a new kind of stuff, merely a novel coexistence of old kinds.

Because there are no chemical reactions involved in the formation of mixtures they can be made without regard to proportions. A good loaf of bread relies on a recipe and the mathematical relationships of its ingredients, much like water requires the combination of exactly two parts of hydrogen with one part of oxygen. A salad, though, can be improvised with whatever odds and ends are lying around in the vegetable crisper. There's no standard formula for a mixture, and no two need be exactly alike. (See Figure I.3.)

In order to decompose a chemically created substance a chemical process must be employed to rearrange atoms and create new varieties of molecules. To break down a loaf of bread, digestive acids must wrench apart complex carbohydrate molecules into molecules of simple sugars. But mixtures, being the products of physical rather than chemical operations, can be separated into their component substances by physical rather than chemical means.

The separation of a mixture into its constituent ingredients takes advantage of the fact that the substances that united to form it retained their physical properties when doing so. Even after being tossed as a salad, carrots remain orange, lettuce green, and tomatoes red. Each can be identified and picked out without resorting to any chemistry.

I.11 FAMILIAR MIXTURES

All the most common occurrences of matter are mixtures. The saltwater that covers nearly three-fourths of the globe is a mixture primarily of water and salt.

Sorting Things Out

The technique used to separate the ingredients of a tossed salad is properly called sorting. Sorting distinguishes among substances based upon how they look. It can be used to pick a needle out of a haystack or nuggets of gold out of a mountain of gravel.

—What are some visual characteristics that can be used in the sorting process?

There are often more efficient alternatives to sorting. Prospectors, for example, usually search for gold by panning. This involves scooping a panful of gravel from a riverbed, swirling it in the current, and hoping a nugget or two shows up at the bottom of the pan after the gravel has been washed away. The method employed in this case is known as gravitation.

—What important, nonvisual, difference between gold and gravel does gravitation rely on?

Flotation works on the same principle as gravitation, only a bit backward. It's commonly used in the dairy business, where cream is collected by allowing it to float to the top of unhomogenized milk and then skimming it off. What's left, of course, is skimmed milk.

—Can you think of any other uses of flotation?

Because common seawater is a mixture it can be separated into its various components by physical means. One way of doing this, called evaporation, yields salt, and another, called distillation, yields pure water. Both of these processes are commonly used where either salt or drinking water is otherwise unavailable.

—How do evaporation and distillation work?

Making coffee involves what is known as extraction to separate the flavorful coffee oils from the bitter grounds. Extraction could also be used to wash the salt out of a mixture of salt and sand.

—What physical characteristic of salt and coffee oils not shared by sand or coffee grounds would such a separation rely on?

There are, of course, many other physical separation techniques. All are based upon one or more differing characteristics among the substances to be isolated. None of them alters the chemical identity of any of the substances involved.

—Could you use physical separation techniques to separate a mixture of crushed ice and glass? How would you do it?

—Can you think of a better way than sorting to find a needle in a haystack?

Beyond these two basic compounds the oceans also contain traces of literally thousands of other substances. The air we breathe is a mixture, too. It's composed of the elements nitrogen and oxygen, the compound carbon dioxide, and a host of minor gases like argon. Like the sea and the sky, the earth beneath our feet is also a mixture. It consists of a variety of rocks, dirts, clays, gravels, and other earthy materials each of which is a different substance or itself a mixture.

Like all mixtures, the sea, the sky, and the earth were created by physical mingling. The sea's mixture of substances is the result of millions of years of evaporation of seawater, its condensation into clouds and then rain, and the extraction and erosion of water-soluble minerals from the earth. The air has, over the ages, become a mixture of all the gaseous substances that escape into the atmosphere. Once airborne these unfettered particles mingle in the global atmospheric ocean. The stuff of terra firma has been mixed by several processes. The denser portions have gradually shifted downward under the force of gravity. Small particles have been tossed about by wind and rain. Everything has been subjected to the agitation caused by volcanoes and earthquakes.

Because these familiar mixtures have been created by physical means they can be also be taken apart without resorting to any sort of chemistry. Seawater is frequently broken down into its component substances by a purely physical process called distillation. Air can also be distilled, but this requires that it first be liquified by means of extreme chilling and high pressure. (See Unit I.23.) Once it's been liquified it can be allowed to slowly warm. As it warms, each of its constituent substances boils off at a different temperature and can be captured as it does so. The distillation of air is an important industry that supplies oxygen to hospitals, nitrogen to fertilizer plants, and other gases, like neon and carbon dioxide, to laboratories and factories for hundreds of uses.

Good old dirt, of course, is the most important source of chemical ingredients. Of the substances that make up the earth's crust, a few are elements and the rest are chemical compounds. Together they're known as minerals and account for everything from mud to molybdenum. Most minerals contain large amounts of oxygen and silicon, the vowels of the chemical alphabet. United chemically these two elements form silica, the major ingredient of common sand. When chemically bonded to a few other elements, they also form quartz and feldspar. Quartz and feldspar, in turn, physically mix to create granite.

Besides mundane elements like oxygen and silicon, minerals may also contain anywhere from a trace to a significant proportion of other more valuable elements. When a mineral's content of a valuable element is quite high the mixture in which it's found is called an ore. Ores, then, are physical mixtures that include minerals containing sought-after ingredients. Lead ore, for example, contains galena, a mineral compound rich in lead. Hematite and magnetite are minerals high in iron content and are present in iron ore. The mineral bauxite is the source of aluminum in aluminum ore. Copper ore generally contains a copper-rich compound called malachite, which is also used as a semiprecious stone.

Because valuable elements like copper, aluminum, and iron are generally embodied within mineral compounds that are then tossed together with all sorts of other minerals, mining them is a two-step process. The first step is to physically dissociate the desirable mineral from the other, relatively worthless minerals. Then, once singled out, the valuable mineral must be broken down chemically to liberate the sought-after element from the other elements to which it's chemically bonded.

The mining of iron is typical. Once an area rich in the compounds of either magnatite or hematite is located the earth there must be physically processed to separate the iron-containing compounds from the other nuisance minerals. This usually involves crushing the mixture, washing it, and sorting it by gravitation—sort of like panning for gold on a large scale. Once the magnatite or hematite has been physically isolated from its worthless fellow minerals it's put in a blast furnace where it's heated along with coke (a form of carbon derived from coal) and limestone. The ensuing chemical reaction rearranges the elements in the furnace to produce carbon dioxide, calcium silicate (also called slag), and iron.

It isn't just the earth's water, air, and crust that exist as mixtures. The food we eat, the people we meet, the clothes we wear, the wood and bricks of our homes, and everything in them are all mixtures. Even the so-called gold in our jewelry and teeth is really a mixture of gold and other metals called an alloy. (See Unit I.20.) Nothing routinely exists in nature in a pure state. The 92 elements have a passion for chemically bonding with one another to form compounds. And compounds, the natural and human-made varieties of which number in the millions, have an equally strong urge to merge physically to form mixtures. (See Figure I.4.)

I.12 ANALYSIS AND SYNTHESIS

It was a bewildering chaos of elements, compounds, and mixtures that mystified early peoples, fascinated the Greeks, defeated the alchemists, and resisted the efforts of the early scientists. Although still a bewildering chaos to most of us, matter is now fairly well understood by a select group of people called chemists. Upon the foundations laid down by Lavoisier, Dalton, and Mendeleev, among others, these matter experts have erected an impressive body of knowledge called chemistry. An inventory of your neighborhood supermarket, your medicine cabinet, or your glove compartment will give you some idea of how far chemistry has brought us from the days of war clubs and reed baskets.

Reduced to its bare essence, chemistry is nothing but the process of taking the 92 different kinds of atoms we have at our disposal—one type for each of the 92 elements—and rearranging them to our liking. It can best be thought of as consisting of two parts: analysis and synthesis.

These elements unite chemically to form	these minerals, which combine physically	to form these common mixtures
magnesium, carbon, and oxygen	magnesium carbonate	limestone
carbon, oxygen, and calcium	calcite	
oxygen and silicon	quartz	
oxygen and silicon	quartz	sandstone
oxygen and calcium	lime	
silicon, aluminum, calcium, and oxygen	feldspar	granite
oxygen and silicon	quartz	
carbon, oxygen, and calcium	calcite	chalk
oxygen and silicon	silica	
oxygen and silicon	quartz	
aluminum and oxygen	bauxite	aluminum ore
oxygen and silicon	silica	
oxygen and iron	hematite	iron ore
oxygen and iron	magnatite	
various	various	
oxygen and calcium	calcite	marble
various	various	
copper, carbon, and oxygen	malachite	copper ore
various	various	
lead and sulphur	galena	lead ore
various	various	
oxygen and silicon	silica	sand
various	various	

Figure 1.4   Elements, minerals, and mixtures

■■■■■■■■■■■■■■■■■■■■■■■■■■■■■■■■■■■■

Chemistry

The study of matter has traditionally been divided into the fields of organic and inorganic chemistry. This distinction was originally based on the long-held belief that some substances contained a special "life force" while others didn't. The former were classified as being organic and the latter as inorganic.

In 1828 a German chemist, Friedrich Wöhler (1800–1882), succeeded in synthesizing urea, previously obtained only from kidneys, out of totally inorganic ingredients. He thereby erased the academic line separating the living from the dead and cast a fatal shadow of doubt on the whole life-force theory.

Partly out of tradition and partly out of convenience, chemistry remains divided into organic and inorganic disciplines. Organic chemistry is now defined as the study of any substance containing carbon. According to this criteria virtually all living as well as many nonliving substances are organic. Such is the nature of carbon that organic compounds far outnumber inorganic ones.

■■■■■■■■■■■■■■■■■■■■■■■■■■■■■■■■■■■■

Analysis is the physical separation of naturally occurring mixtures into their various chemical substances and the subsequent chemical reduction of those substances into useful elements and compounds. Analysis provides the inventory of raw materials that are needed for the subsequent construction of custom-made complex substances. The earth's crust yields iron, copper, and lead; coal yields carbon; water yields hydrogen and oxygen; and the air yields nitrogen via analysis. Just about everything a chemist needs can be found in commonplace, often abundant, materials. Theoretically, every one of the 92 elements can be coaxed out of nothing more than plain old seawater.

Once a stockpile of basic ingredients has been built up via analysis, synthesis begins. This is the creative part, an orchestrated restructuring of the indestructible raw materials into new formats. Artificial materials—those made intentionally under human supervision—now outnumber naturally occurring ones, and in many instances are used in place of them because of their superior performance. Carbide-steel drill bits have all but supplanted flint chisels, fiberglass fishing rods have made their bamboo forerunners pretty much obsolete, Melmac dinnerware is much preferred to pieces of bark, and floors of packed dirt have become rather undesirable when compared to the durability and convenience of linoleum. The immediate living environment of *Homo sapiens* is being reshaped according to its own specifications.

The processing of raw petroleum is a classic example of how chemistry transforms a common and relatively useless natural material into a virtual cornucopia of elaborate, exotic, and valuable synthetic products. As it comes from the ground, crude oil is a smelly, dirty, and random mixture of many chemical compounds, most of them rich in hydrogen and carbon and thus called hydrocarbons. By the mechanical process of fractional distillation—a repeated series of boilings and condensations—this hydrocarbon stew is physically sorted out into its basic chemical ingredients. Some of the compounds thus isolated are immediately useful and require no further processing. Ethane and methane, for example, are ready to use as natural gas, a clean-burning and highly efficient fuel. Propane and butane are liquified under pressure, bottled, and also used as a fuel. They're known, either individually or collectively, as liquified petroleum, or LP gas.

Most of the hydrocarbons sorted out from raw petroleum, however, are large, clumsy, and relatively worthless in their natural state. These complex substances must be chemically processed to make them valuable. The first step in this chemical processing is one of analysis. By a procedure known as cracking, the cumbersome molecules are fragmented into many smaller molecules, and the basic hydrogen and carbon atoms are thus made more accessible. The cracking process relies on high temperatures, high pressures, and the presence of catalysts, substances that, like the promoter of a prizefight, facilitate chemical reactions but don't directly participate in them.

The smaller molecules that result from the cracking process can then be selectively pieced back together in desirable formats to form gasoline, kerosene, motor oil, diesel fuel, wax, and asphalt. Pressure and heat are again the primary tools used. When ingredients from sources other than petroleum are introduced into the equation the list of potential products mushrooms to include plastics, rubbers, detergents, synthetic fibers, drugs, cosmetics, and insecticides, among many others.

It needs to be pointed out that this sort of chemical magic, although wonderful, isn't without its messier aspects. It can, for example, be excruciatingly long and tedious. To get from a raw material to a finished product may require hundreds of delicate operations and involve a host of intermediate substances. It's frequently very inefficient as well. A carload of ingredients might be needed to produce a cupful of a particular end product. It can also be downright hazardous. Most chemical operations generate many unwanted by-products along the way. Some of these can be further processed into useful materials, but others are either worthless nuisances that must be disposed of or, much worse, dangerous toxins harmful to the environment and its inhabitants.

It's all done in the name of progress—the search for a redder tomato, a more wrinkle-resistant shirt, or more luxurious hair. Theory, analysis, and synthesis are only means to an end. In some ways we're like the ancient Greeks—we have an innate curiosity about matter and how it works. But mostly we're like the alchemists. We want to turn lead into gold.

I.13 | TWO UNIVERSAL PROPERTIES

Just what is it that makes gold so much more desirable than lead, anyway? What makes the rather involved transformation of petroleum into asphalt worth all the trouble? Does anyone really care if the atoms in their wedding band have an atomic number of 79 rather than 82? Does anyone know what kinds of molecules cover their driveway? Of course not. The inner structure and composition of matter is of little interest to most of us. We differentiate, assess, and utilize matter on the basis of its outward characteristics—those we can taste, smell, see, and feel.

From a scientific standpoint the discussion, description, and evaluation of these characteristics nearly always involves three steps: an unequivocable definition, an objective measurement, and a numerical expression of the results. Obviously, not all characteristics can be fitted into this rigid formula. The Greeks found out a long time ago that beauty is impossible to define. To measure how delicious something is or numerically express comfort is equally hopeless. Yet there are plenty of material attributes that lend themselves readily to this treatment. There are two in particular that almost demand it.

Matter, you may recall, is usually defined as anything that has mass and occupies space. The possession of mass and the occupation of space, then, are the two fundamental characteristics shared by all matter. Both of these universal properties are relatively easy to define, measure, and numerically express.

Let's start with the occupation of space, the physical property usually called volume. Volume can be defined as the size or extent, in three dimensions, of any sample of matter, be it an atom or an alp. Any such sample, according to the definition of matter, will have a volume, and this volume can be objectively measured and numerically expressed according to any of several different systems. Pints, gallons, cubic meters, and bushels are all measurements of volume that simply describe how big something is—how much space it occupies.

The second universal property of matter, mass, is the amount rather than the size of any given sample of matter. Any sample of matter, be it a molecule, a mountain, or a molehill, will have a particular mass just as surely as it will have a particular volume. And this mass, like its volume, can be impartially measured and numerically stated. A measurement of mass, be it tons or kilograms, simply describes how much of something there is.

Volume and mass, being possessed by all matter, are two concepts with which we're quite familiar. We use them all the time to discriminate between objects and materials. A softball and a baseball are outwardly virtually identical except for their size, or volume, and are easily identified on this basis. A full and an empty can of hair spray differ only in the amount of matter they include. This difference in mass is sufficient to tell one from the other. Whether we're aware of it or not, our concepts of nearly any and every bit of matter we encounter include estimates of size and amount.

Because volume and mass are so utterly familiar, they're also quite well understood. Descriptors of size and amount can be used with confidence to describe almost anything without fear of misunderstanding or ambiguity. There is, however, one twist in this area that deserves some special attention. It concerns the difference between what is technically referred to as mass and what is commonly referred to as weight. The two, often confused or used interchangeably, aren't exactly the same.

Mass is a measure of the quantity of matter present in any sample of stuff. Weight is the measure of a sample of stuff's attraction to the planet Earth under the influence of gravity. An object's mass is an intrinsic and invariable physical attribute. Its weight isn't. An astronaut sitting on the launchpad at Cape Canaveral has weight. The same astronaut, circling in orbit a few hours later, is weightless. The astronaut's mass, meanwhile, has remained constant. Mass is therefore a reliable and constant descriptor of a chunk of matter while weight is merely a temporary condition dependent upon position.

Seeing as how most of our time is spent on earth, where weight doesn't appreciably change, weight as well as mass can generally be used to describe bulk. The two terms can, and almost always are, used as equals in common situations. It's good to remember, though, that there's a difference. A helium-filled balloon, floating weightlessly, has a positive and very real mass. Everything does.

$\boxed{\text{I.14}}$ DENSITY

Mass and volume, although universal attributes of matter, still leave something to be desired as descriptors and identifiers. In fact, it's because absolutely every bit of matter has both a volume and a mass that these two properties are, in and of themselves, fairly useless.

A spoonful, a quart, or a cubic centimeter of matter could be virtually anything from cough syrup to dry-cleaning fluid. All matter has volume and any given volume can be associated with any substance. Likewise, a 20-gram, 200-ton, or 2-ounce container of matter could be loaded with sand or Silly-Putty. All matter has mass and any given mass can be associated with any substance. Properties like mass and volume, which can be imparted to a number of different substances or can assume a wide range of values for any given substance, are called accidental properties. These properties, which also include shape and temperature, relate more to a temporary condition of matter than to its permanent nature. Clearly, then, it's not accidental properties that make gold more esteemed in cuff links than lead or asphalt more appropriate for a driveway than sugar.

The attributes that help us discriminate among substances and set one apart from another for a particular use are called specific properties. Specific properties are an intrinsic and unalterable part of a substance's identity. A lump of coal may assume nearly any shape, size, or bulk, but it will forever retain its combustibility. A piece of copper wire, no matter what its length or temperature, will always be reddish-yellow and a good conductor of electricity. It's by the use of specific properties that we deal with the material world. We know that something black that burns might well be coal and most definitely isn't copper.

Just as all matter possesses certain accidental properties like volume and mass, so it also possesses certain specific properties. One such universal specific property is density. Because every bit of matter has both a mass and a volume it will also have a density, for density is the ratio of these two accidental properties. Density is expressed as units of mass per unit of volume: pounds per cubic foot, grams per cubic centimeter, or any other such combination. Because it's a property of all matter, easily determined, and extremely reliable as an identifier, density is probably the most useful of all specific properties.

We all use density to sort out the substances that surround us, whether we realize it or not. It's usually density, and not weight, with which it's often confused, to which we're referring when we discuss something's heaviness or lightness. It's an easy enough mistake to make. Aristotle was pretty mixed up himself when he assigned gravity and levity to all things Greek.

We, like Aristotle, are inclined to say that lead is heavier than cork, but this declaration of weight, by itself, is pretty meaningless. Is a kilogram of lead heavier than a kilogram of cork? Of course not. Both weigh precisely one kilogram and their weights are thus equal. Is the cork from last night's bottle of champagne lighter or heavier than the lead sinker in Dad's tacklebox? Until they're both weighed it's impossible to say. What we *can* say is that lead is denser than cork. A given volume of lead will be heavier than an equal volume of cork. To be precise, the density of lead is 11.3 grams per cubic centimeter while that of cork is a mere .22 grams per cubic centimeter. While a 10-gram sample of matter might be lead, cork, lithium, or cookie dough, a sample with a density of 11.3 grams per cubic centimeter is more than likely lead. A sample with any other density surely isn't.

For the sake of convenience density is sometimes measured and expressed as specific gravity. Specific gravity is the ratio of a substance's density to that of pure water. The specific gravity of water is therefore 1.000, as is the ratio of anything to itself. The specific gravity of ice turns out to be .917. Ice, because its density is less than that of water, has a specific gravity of less than 1.000. The specific gravity of seawater is 1.025. This tells us that the density of seawater is fractionally greater than that of pure water. Anything with a specific gravity of less than 1.000 has a density less than that of water and will float in it. Anything with a specific gravity greater than 1.000 is denser than water and sinks in it.

Anything, in fact, will float in a fluid with a specific gravity or density greater than its own and sink in a fluid with a specific gravity or density less than its own. The human body, with a specific gravity of 1.015, sinks in 1.000

• •

A Famous Bath

King Hiero of Syracuse, on the island of Sicily, had a problem. He had just taken possession of a new crown made by a local goldsmith and he wasn't sure he liked it. The king didn't trust the goldsmith, you see, and suspected the man may have cheated him by using some silver or copper in the crown instead of nothing but gold as instructed. But how to prove it?

As he generally did when in trouble, the king called upon his loyal subject and advisor, Archimedes. And, although he had no idea just how he would do it, the brilliant philosopher agreed to help. Certainly he couldn't melt down, break open, or harm the crown in any way.

Archimedes was no metallurgist, but he knew that one of gold's outstanding characteristics was its density. Gold, in fact, was denser than any other known metal that could have been used to dilute the crown's purity. The problem, then, became finding the density of the crown and comparing it to that of pure gold.

Being an accomplished mathematician, Archimedes knew well that the density of any object was determined simply by dividing its weight by its volume. Weighing the crown presented no special problem—it could just be put on a scale like a bunch of grapes or a goat's bladder of olive oil. Finding the volume of the crown, though, was quite another thing. Because of its elaborate shape there was no way to measure it with a ruler or calculate its size with a formula.

Archimedes was stuck. The entire problem of determining the crown's purity was reduced to finding its volume and he knew of no way to do it. It was at this point he decided to take a bath. As he stepped into the tub he noticed how the water level rose and knew his problem was solved. Elated, he leapt out of the bath and ran through the streets shouting "Eureka!" (I've found it!).

What Archimedes found was the principle of measuring volume by liquid displacement.

—How does it work? Think of dropping a set of keys into a measuring cup that is partially filled with water.

—Archimedes determined that the goldsmith had, indeed, tried to cheat the king. Was he able to figure what other metals the man used? What about the quantity of other metals?

—Is there any way that a clever goldsmith could have fooled Archimedes? What about one with access to metals denser than gold?

• •

pure water and floats in 1.025 seawater. Likewise helium, with a specific gravity of .0002, "floats" on air, rated at .0013. That's why a helium-filled balloon, despite a very real mass, seems weightless and drifts skyward when released.

Minor differences in specific gravities can have major ramifications. Salt water is slightly denser than fresh water; cold water slightly denser than warm. These two facts work together to account for most of the world's ocean currents and accordingly affect not only shipping and fishing, but global weather patterns as well. Cold air, too, is slightly denser than its warmer counterpart. Wind, cold fronts, warm fronts, high-pressure areas, low-pressure areas, and all they produce are the direct results of the dynamics that take place between masses of air with different specific gravities. (See Units I.24 and II.14.)

I.15 A MULTITUDE OF CHARACTERISTICS

Density is but one of a multitude of characteristics that differentiate one substance from another. There's also color, texture, aroma, hardness, flexibility, reflectivity, transparency, elasticity, firmness, strength, and many, many others. Some of these may be considered crucial, some simply interesting, and still others only faintly amusing or totally inconsequential depending on the context in which they appear and the person making the judgment. When it comes to the operation of the physical world, though, the evaluation is a bit more straightforward.

Among those considered critically important, after density, is something called heat capacity. Heat capacity is, perhaps rather obviously, the measure of a substance's ability to absorb heat. Substances with a high heat capacity are capable of soaking up all sorts of heat without becoming much hotter. Substances with low heat capacities warm up quickly when even a little heat is added to them. For the sake of convenience heat capacity is usually discussed in terms of specific heat. Just as specific gravity is the ratio of a given substance's density to that of water, so specific heat is the ratio of a substance's ability to absorb heat to that of water.

The ability to absorb heat differs greatly from one substance to the next. Water, because it's used as the standard against which other substances are measured, has a specific heat of 1.000, the ratio of anything compared to itself. The specific heat of iron, by comparison, is .107, that of copper .090, and that of lead or gold a mere .031.

Water, as the above numbers indicate, is a spectacular performer when it comes to absorbing heat. Water heats up very little when heat is added to it or, to put it another way, it takes a great deal of heat to raise the temperature of water. Based on their specific heats, we can see that it takes over 30 times as much heat to warm up a specified amount of water from 40 degrees, say, to 50 degrees than it does to warm an equal mass of lead or gold the same 10 degrees. Conversely, a sample of water that cools off by 10 degrees will give off more than 30 times as much heat as an equal mass of lead or gold that is cooled by a similar amount. Whatever heat goes in to raise the temperature comes out when the temperature drops.

The outstanding ability of water to absorb and then liberate heat makes the world a more comfortable place to call home. The oceans absorb enormous amounts of heat in the form of solar energy during the day, then release them at night, thereby providing us with cooler days and warmer nights than we would otherwise experience. The water present in the air as humidity works the same way. The same stabilizing mechanism, operating over a longer time frame, helps keep summers a bit cooler and winters a bit less harsh than they would otherwise be. Dry, landlocked deserts provide local examples of what extreme temperatures can result in the absence of water's moderating influence.

Water's great heat capacity also has small-scale applications. An automobile's engine is kept from overheating by a water-filled cooling system. The water soaks up vast amounts of heat as it circulates around the engine, then releases this heat to the surrounding air as it passes through the radiator. Water performs a similar function in many heating systems. In this case the water is loaded up with heat by a furnace, then pumped to outlying radiators where the heat is released and a building warmed. (See Unit II.15.) What serves as a blessing can also be a curse, however. Anyone who has waited impatiently for a coffeepot to perk or a bowl of soup to cool has the exceptional heat capacity of water to blame.

Hardness is a less important but more commonly considered attribute of matter. Technically speaking, hardness measures a substance's ability to scratch or its resistance to being scratched. Like mass, volume, density, and heat capacity, hardness can be objectively measured and numerically described. A hardness scale has been developed to do just that. In this case the standard against which other substances are measured is diamond, not water. Diamond, the hardest known natural substance, rates a 10.00 on the hardness scale. Rubies and sapphires, with hardness rankings of 9.0, aren't quite so hard. They can be scratched by a diamond but not by glass, which is rated at 5.5. A 3.0 copper penny will scratch a 2.0 fingernail, but will be scratched by either a 5.0 pocketknife blade or a 6.5 steel file.

Solubility measures a substance's tendency to dissolve. It varies for each substance according to the solvent involved. (See Unit I.25.) Water, not surprisingly, is the most common solvent. Sugar dissolves in water to the extent of about 70 parts of sugar per 100 parts of water. Salt dissolves to the extent of about 26 parts of salt per 100 parts of water. Iodine is insoluble in water, but dissolves readily in alcohol, another popular solvent. Gasoline dissolves grease, which is insoluble in either water or alcohol.

■■■■■■■■■■■■■■■■■■■■■■■■■■■■■■■■■■■

Diamond

Diamond is the hardest substance found in nature, but a human-made compound of boron called borazon is even harder. Although regarded as being neither especially beautiful nor prestigious, borazon is valuable in industrial applications where it can be used to grind, cut, or scratch virtually anything.

Diamonds, too, are employed in many mundane ways, mainly in abrasives, drill bits, and other tools. For economic reasons the diamonds used in this context are usually manufactured rather than mined. Although they're commonly called artificial, chemically speaking these stones are every bit as genuine as their natural counterparts.

With enough heat and pressure diamonds can theoretically be made out of anything that is rich in carbon, including peanut butter. Coal is more commonly employed. Regardless of how or out of what they're made, however, manufactured diamonds are all small and of very poor quality due to impurities. Neither they nor borazon seems likely to replace natural diamond as a birthstone, an engagement ring, or a symbol of value and reliability.

■■■■■■■■■■■■■■■■■■■■■■■■■■■■■■■■■■■

Malleability measures the ease with which a substance's shape can be altered. A penny can be hammered into a flat disk; a piece of iron can't. Brittleness, a sort of inverse to malleability, measures a substance's resistance to being distorted. Steel flexes under stress; china shatters. Elasticity measures the tendency of a distorted substance to resume its original shape. A squeezed rubber ball resumes its roundness when the pressure is released while a lump of clay remains distorted by a similar force. Conductivity is the capacity to conduct an electric current. Resistivity is the capacity to inhibit such a flow. The copper in an electric wire, like most metals, is a good conductor. The plastic sleeve surrounding it, called an insulator, is highly resistant. A substance's melting point is the temperature at which it turns from a solid to a liquid. Its boiling point is the temperature at which it turns from a liquid to a gas.

These physical properties, along with many others, define stuff in a practical sense. Characteristics we can detect and evaluate with our senses tell us what something is, what it can be used for, and what it's worth. They allow us to tell the difference between salt and sand. They make concrete better than clay for building a bridge. They make it easy to distinguish rubies from rubber. They make gold more valuable than lead.

I.16 SUBMICROSCOPIC GOINGS-ON

The incredible variety of physical properties with which the stuff around us is endowed is nothing short of astounding. But, thanks to the labors of Lavoisier, Dalton, Mendeleev, and their successors, this dazzling display of diversity is no longer a source of superstition, speculation, or even mystery. We no longer need to wonder why water is wet. We know. We know, too, why earth is denser than air. We know why gold glitters and lead doesn't. We know what makes steel flexible, china brittle, rubber resilient, copper conductive, and diamond hard. We have the answers that the ancients never even dreamed of, the Greeks could only guess at, and the alchemists simply assumed.

These answers, without exception, lie buried deep within the molecular structure of matter. Submicroscopic goings-on explain why we experience what we do on a macroscopic scale. These invisible events can all be understood with the help of a bit of imagination and the kinetic molecular theory of matter.

The kinetic molecular theory rests on two basic assumptions, both by now convincingly documented and commonly accepted. The first is that, like Democritus supposed, all matter consists of tiny, individual pieces. The second is that each of these specks is in perpetual motion. Every molecule, according to this theory, is like a little nonstop jitterbug, and it's this ceaseless molecular dance that makes matter such dynamic, potent, and fascinating stuff.

The universal properties of matter are easily explained in molecular terms. The mass of any sample of matter is merely equal to the sum of the masses of its individual molecules. The volume of a sample of matter is likewise equal to all the space occupied by its component particles. And density, much as Boyle surmised, is the result of how tightly or loosely packed any particular type of matter's molecules happen to be. The molecules in a slug of lead are lined up shoulder to shoulder, head to toe, and back to belly while those in a piece of cork are spread apart and have empty spaces among them. Lead accordingly has more particles and more mass in any given volume of space. When the empty spaces are eliminated the density of a substance depends to a large degree upon the weight of its constituent particles. It's no mere coincidence that an atom of lead weighs just about 11.5 times as much as a molecule of water and that the density of lead is 11.5 times that of water. Indeed, it was relationships like this that allowed Dalton to come up with his table of atomic weights without ever having isolated a single atom.

Heat capacity is also the result of how matter is constructed at the molecular level. For reasons that are too complicated to explain here and imperfectly understood in any event, it seems that the ability of a substance to absorb heat is inversely proportional to the mass of its component particles. Water, with extremely light molecules, is an outstanding absorber of heat. Iron, with medium-weight particles, is an average absorber, and lead and gold, with heavy particles, are poor heat absorbers.

Hardness also depends on the molecular structure of matter. The two allotropes of carbon—pencil lead, or graphite, and diamond—provide a classic illustration. Graphite, remember, is constructed more or less like a stack of graph paper. It consists of two-dimensional sheets layered one on top of the other. This is what makes graphite fairly soft and ideal as a writing medium. The sheets slip and slide over one another and finally peel off as a pencil is dragged across a sheet of paper, thereby easing its movement and leaving a visible trail. A diamond, meanwhile, consists of carbon atoms connected in all three dimensions. In chemistry, just as in architecture, this arrangement results in great rigidity. The Eiffel Tower is sturdy and diamond is hard. (See also Unit I.8.)

Solubility depends upon the relative strengths of intermolecular affinities rather than upon their group geometry. If the attractions among the particles of a solute (the substance to be dissolved) are stronger than those between the molecules of the solute and those of the solvent, nothing happens. The solute retains its integrity as its molecules cling to one another. But, if the solute's molecules have a greater affinity for the solvent's molecules than for one another, the solute disintegrates. A powerful solvent literally dismantles a solute by taking it apart molecule by molecule. The molecules in a grain of sand stick together in defiance of water's efforts to separate them. Those in a grain of salt, however, succumb to the outside tug.

Malleability results when a substance's component parts can be rearranged without being separated. The molecules in a penny slither over and around one another when struck by a hammer, but still maintain contact. Those in a porcelain teacup either hold fast to one another or let go completely, depending on the force of the blow.

Elasticity on a visible scale is nothing more than the bending, stretching, and recoiling of zillions of invisible intermolecular attachments. Conductivity and resistivity measure the ease or difficulty with which tiny bits of electricity, called electrons, are handed along from one molecule to the next. (See Unit II.28.)

I.17 THREE PECULIAR PERSONALITIES

The most graphic illustration of the kinetic molecular nature of matter takes place at boiling and melting points. Here the effects can be appreciated without any sort of measurement, quantification, or instrumentation. The differences between ice and water and between water and steam aren't only obvious, they're dramatic. Yet ice, water, and steam are all the very same chemical compound. Each consists of the very same sort of molecules. The paradox is explainable in terms of molecular motion.

Water, it just so happens, provides the ideal illustration of the three states, or phases, of matter: solid, liquid, and gaseous. Not only is this marvelous stuff totally familiar, but its three peculiar personalities all occur at temperatures well within the range of everyday experience. In this sense it's unique.

Virtually every chemical compound known, however, is theoretically capable of existing in any and all of the three states of matter. Naturally occurring gases, like the oxygen and nitrogen in the atmosphere, as well as the air itself, can be liquified when cooled sufficiently. Further and extreme cooling will actually turn these substances into solids. A few gases, like hydrogen and helium, can only be liquified with a combination of exceptionally low temperatures and exceptionally high pressures. An extra squeeze is also needed to liquify carbon dioxide. At regular atmospheric pressure this common gas goes directly from the gaseous to the solid state when cooled, bypassing the usual intermediate liquid phase. Solid carbon dioxide, better known as dry ice, similarly goes directly into the gaseous format as it warms. This leaping over the liquid transition is called sublimation.

Naturally occurring solids, conversely, can, with a bit of persuasion, be forced to assume liquid or gaseous dispositions. Normally solid rock is liquified in a volcano and normally solid iron is liquified in a blast furnace. Higher, often seemingly impossibly high, temperatures will eventually vaporize these and all other solids as well. The sun is a ball of gases, most of which would be solids under less extreme conditions.

The transition from solid to liquid, from liquid to gas, and then back again is a purely physical change, not a chemical one. Each molecule remains intact. What changes is the energy level and organization of the molecules as a group. (See Figure I.5.)

In the solid state molecules are arranged in fixed positions relative to one another, held in place by a mutual affinity that overpowers their individual kinetic energies. Like a classroom of third-graders sitting at their desks, they wriggle and squirm, but their location among their fellows doesn't change. A solid has an unchanging volume, density, and shape because its constituent molecules are interlocked in a definite pattern. A block of ice is cohesive, rigid, and needs no container to hold it because its internal structure is strong enough to withstand external forces such as the downward tug of gravity.

In a liquid molecules don't maintain any sort of position or order, but roam freely among one another. This relative mayhem, however, takes place with each molecule still attracted to the rest. Like third-graders when the teacher is out of the classroom, liquid molecules mill about freely with regard to each other, but retain their identity as a group. Their individual energy levels are higher than those of their solid counterparts, but still not high enough to overcome their adherence to one another. They're out of their desks and jumping around, but still confined within the walls of the classroom. It's this condition that lends to liquids a definite volume and density but no definite shape. Internal attractions are sufficient to keep molecules together in a cohesive clump, but not strong enough to defy gravity. A cupful of water

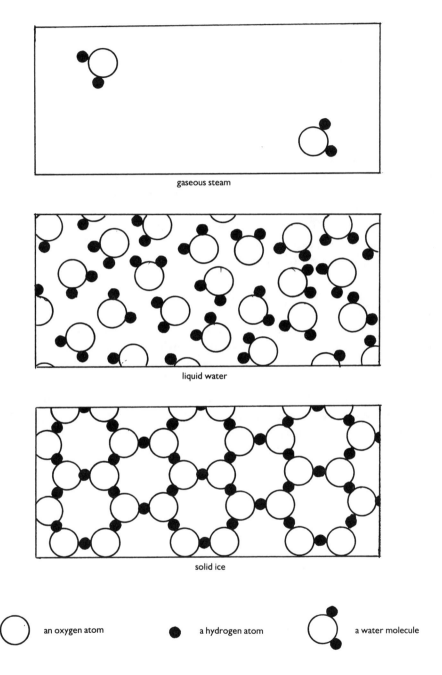

Figure I.5   The three states of matter

remains a cupful of water regardless of the container in which it is placed, but this restless mass has no shape of its own.

The gaseous state, finally, pays no heed at all to any sort of constraint or organization. The molecules within a gas act pretty much as they wish. With individual energies that overcome any sort of mutual attraction, gas molecules behave as a collection of autonomous units, not as a united group. Like third-graders after the final bell, they take off in their separate directions free of any group discipline. Unless contained on all sides the molecules of a gas will simply wander off. Gases, such as steam, have neither a certain volume, density, nor form. They'll fill any container regardless of its size or shape.

The existence of matter in solid, liquid, and gaseous states is everywhere evident. The world we inhabit, in fact, can accurately be thought of as a great, three-layered ball with one layer for each of matter's three phases. At the core of this massive ball is a solid layer called the lithosphere. It's composed of the rocky continents and the seabeds. The middle layer is a liquid one called the hydrosphere. It consists of the oceans, lakes, rivers, and other waterways that cover the globe. The outermost layer is the atmosphere, the gaseous blanket that shields us from uninhabitable outer space.

I.18 SOLID STUFF

Perhaps because we're land-dwelling creatures, captives of the lithosphere, we're generally most comfortable with the solid state of matter. Solid stuff is tangible, predictable, firm, stationary, and seemingly permanent. Solids, as a rule, don't evaporate, run down the drain, or blow away. As might be expected, though, there's more to solids than meets the eye. Their true characteristics depend not upon their outward appearances, but upon their inner structures.

There are, to begin with, two basic and quite distinctive types of solids: the vitreous or amorphic type and the crystalline type. Amorphous solids, like wax or glass, consist of molecules arranged irregularly or randomly within a solid clump. The links among these molecules are fairly rigid, but follow no pattern. Some linkages are longer than others, some shorter. Some strong, some weak. Crystalline solids like gems and minerals, in contrast, consist of molecules stacked with geometric precision. Links among molecules are all equal in length and strength and aligned in specific directions.

Amorphous solids, reflecting an inner lack of discipline, behave less regularly than do crystalline solids. Whereas crystalline solids turn suddenly to liquids at temperatures that can be measured to a fraction of a degree, wax, glass, and their vitreous fellows gradually soften. Whereas crystals tend to break along clean, sharp lines, amorphous masses tend to stretch or shatter. Because all the intermolecular links within a crystal are identical they all behave in unison. Melting takes place everywhere at once. Because the intermolecular links in vitreous solids are all somewhat different they each behave in their own peculiar manner, breaking at different stress levels and in different ways.

Amorphous solids are really less solid than their crystalline counterparts. They lack many of the classic characteristics that define the solid state. For this reason they're sometimes referred to not as true solids, but as supercooled liquids. They even display some liquidlike properties under the right conditions. Glass, for example, will actually distort or "flow" when subjected to a long-enough and strong-enough pressure. For the most part, though, this is a mere technicality. New types of glass have been formulated that resist pressure as well as any true solid. Glass, in fact, because it's durable, moldable, transparent, nonporous, nontoxic, rustproof, and made basically out of nothing more than silica sand, is rapidly becoming a modern supersubstance. The burgeoning varieties, characteristics, and applications of glass are probably exceeded only by those of another well-known amorphous solid—plastic.

It's hard to overstate the importance of plastic. Had it been known to the ancient Greeks it probably would have been included as one of the original four elements or perhaps added as a fifth. The development of this amazing stuff, able to take on the guises of wood, glass, china, cloth, rubber, gems, cardboard, and leather, as well as assume identities totally its own, is nothing less than the story of modern chemistry.

The first plastic was produced, more or less by chance, by an American inventor, John Wesley Hyatt (1837–1920), in 1868. He and his brother Isaiah were looking for a cheap substitute for ivory in the manufacture of billiard balls when they cooked up a batch of cotton cellulose and camphor. The result was something they called celluloid. It made a serviceable billiard ball and soon found other uses as well, most notably as photographic film.

Celluloid was the prototype of a number of early plastics, all of which were made out of completely natural ingredients combined in completely unnatural ways. The first wholly synthetic plastic, made out of ingredients that were themselves the result of a contrived chemical process, wasn't formulated until 1906. That's the year that Leo H. Baekeland (1863–1944), a Belgian-born American, invented and named after himself Bakelite, a hard, nearly indestructible material that found immediate applications in the then fledgling automobile industry.

These early plastics were nothing more than the results of lucky accidents via the process of trial and error. It wasn't until the 1920s that plastics really came into their own when natural cellulose, the stuff of which wood and celluloid is made, was found to consist of long chains of simple sugar molecules. Mimicking and improving upon nature, chemists began stringing together all sorts of simple compounds into long, multimolecular necklaces. This process, called polymerization, lies at the very heart of the plastics industry and has made it an all but irresistible force since its perfection.

Despite their incredible range of physical properties, plastics all share the same basic chemical structure—all are polymers. Everything from toothbrushes to Teflon consists of huge macromolecules—chains of thousands of tiny molecular fragments linked end to end in a repeating pattern. The main

● ●

Abracadabra!

Imagine you are a chemist working in a plastics factory. You've studied the theory of atoms and molecules for years and know the periodic chart as well as the alphabet. You've spent endless hours in laboratories studying the behavior of elements and compounds, and you know how to use heat and pressure to make them perform in seemingly magical ways.

You want to create a plastic that is, above all, extremely strong. You don't much care, for example, what it looks like. As always, you want to keep its cost down. That means it should be fairly easy to manufacture and made out of common ingredients.

The first thing you do is round up some nitrogen and water. These you chemically combine to produce ammonia. The ammonia you chemically alter to form a substance called hexamethylenediamine. Its seemingly nonsensical name, like those of many compounds, is derived from its elemental composition and the architecture of its molecules.

Next you use distillation to collect some carbon-rich benzene from a sample of natural gas. The benzene, in turn, you transform into a synthetic compound called adipic acid.

You're ready now to form a polymer. You take a molecule of adipic acid and join it to a molecule of hexamethylenediamine. To this double molecule you attach another of adipic acid, then one of hexamethylenediamine, and so forth and so on, like someone stringing a garland of cranberries and popcorn for their Christmas tree.

The resultant macromolecule is just what you want. It can be strung out into filaments and used as fishing line. A number of filaments can also be braided to make rope. Fine strands of it can be tightly woven to form a lightweight, wind-resistant fabric perfect for camping gear. A looser weave results in a sheer fabric used in women's stockings.

What you've created, of course, is called nylon. It's only one of hundreds of plastics you can make to order for special applications.

—Where and how do you get nitrogen? What common substances would you look to as sources of hydrogen, silicon, and oxygen?

—Styrofoam, another of your products, is used to make disposable coffee cups, ice chests, and life preservers. What properties make it desirable for each of these uses?

—What properties would you want in a plastic for making dashboards for cars? Dolls? Sunglasses? Polyester fabrics?

● ●

■■■■■■■■■■■■■■■■■■■■■■■■■■■■■■■■■■

Teflon

Teflon, officially known as polytetrafluorethylene, is the happy result of a 1938 laboratory accident. Because it had exceedingly important and highly sensitive military applications it remained a closely guarded secret for many years. Its use for frying eggs should be considered a mere by-product of less benign activities.

Besides being slippery, Teflon can be stretched into incredibly thin membranes. The membranes are full of millions of tiny holes and can be used as extremely fine filters. The developers of the atomic bomb used Teflon membranes to sift uranium, separating it molecule by molecule into its various isotopes, some of which would support a chain reaction and others that wouldn't. (See Unit II.31.)

Teflon membranes are now popular in waterproof clothing. Their holes are of such a size that they repel water in its liquid form but permit the passage of the gaseous molecules of water vapor. A raincoat using such a membrane keeps the wearer dry both by preventing rain from coming in and by allowing perspiration to evaporate and escape.

■■■■■■■■■■■■■■■■■■■■■■■■■■■■■■■■■■

ingredient in these chains is usually the element carbon. To a long central spine of carbon atoms can be attached other elemental atoms mined from seawater, limestone, or the air. Every new combination results in a new plastic.

I.19 CRYSTALLINE STRUCTURE

Solids originating in nature, in contrast to those originating in factories, are generally crystalline rather than amorphous. Given the opportunity, most collections of identical or very similar molecules tend to organize themselves into neat, three-dimensional rows and columns with military-like precision. These highly ordered formations are crystals. The number of participating molecules needed to form a crystal of visible size is staggeringly large. The likelihood of so many compatible molecules existing together is conversely staggeringly small. Accordingly, the occurrence of identifiable crystals like diamonds and rubies is extremely rare. Instead, the great masses of crystals out of which natural solids are composed are so small they escape notice by the naked eye. They are, nevertheless, true crystals and possess properties that are uniquely crystalline.

Again, it's the molecular construction of crystals, not their outward appearance, that governs their behavior. Chemists, in an effort to organize, study, and understand crystalline structure, have divided crystals into over 30 distinct varieties depending upon their architecture. These varieties all fall into six major categories, or systems, each characterized by its own proportions and shape. There are, for example, crystals that are perfect little cubes—all equal sides and all right angles. These are the most common type. Diamond, gold, silver, and salt all exist in molecular patterns that resemble little dice. A close inspection of a grain of salt will bear this out on a visible scale.

Tetragonal crystals, meanwhile, are shaped like miniature boxes of saltines: square ends, rectangular sides, and all right angles. Tin and zircon exist in this shape. Orthorhombic crystals, like those of topaz and iodine, are rectangular on all sides and look like miniature bricks. The monoclinic crystals of sugar and gypsum are like tetragonal ones but distorted in one direction, while the triclinic crystals of boric acid are distorted in all directions. Calcium, quartz, and ice exist as hexagonal crystals. When water crystallizes in two dimensions the result is also six-sided, as the scrutiny of a snowflake will reveal.

Crystals in sizes big enough to see with the naked eye, while rather rare, are obviously not unknown. Salt and sugar granules are household examples. Sand exists as visible crystals of any of several minerals, most commonly silica, a chemical compound of silicon and oxygen. Larger crystals of silica are known as agate and quartz, smaller ones as cement. Diamonds are visible crystals of pure carbon, rubies and sapphires are made out of aluminum oxide, emeralds out of a compound of berylium, and garnets and zircons out of various silicon compounds.

Because all these substances are crystals they all share certain tendencies. Because their internal forms are so precise their outward behavior is also quite striking. Many crystals, most notably the gems, are treasured for the sheer beauty of their exactness and perfection. Besides being geometrically ideal, they're extremely hard and resistant to change. Their perfect alignment of molecules is also what makes them reflect and refract light with such dazzling efficiency. Their hardness also makes them useful in more practical ways. Low-quality diamonds are used industrially as abrasives, and rubies are the tiny jewels used as wear-resistant pivot points in fine watches. Quartz crystals are found in other watches, where their unerringly accurate vibration patterns mark off fractions of seconds. (See Unit IV.12.) The accuracy and dependability of crystals are also put to work keeping radio transmitters tuned to the correct frequency and in a host of electronic applications including many computers.

Other crystals have more mundane but equally important functions. Sand and cement, two crystalline forms of silica, are basic building blocks of much construction. Other structurally important crystals include those of sandstone, limestone, chalk, marble, gypsum, dolomite, meerschaum, talc, and asbestos. Mica, feldspar, and clay consist of crystals of silicon and aluminum compounds. When fired, these become ceramics, bricks, pottery, china, and porcelain. Nearly the entire lithosphere, in fact, is composed of crystals of one kind or another.

I.20 A METALLIC MEDLEY

The lithosphere is a massive conglomeration of crystalline solids. Almost without exception the crystals of which it's composed are made out of oxygen and a medley of metallic elements. It's been our ability to isolate and utilize these metals that has measured our rise out of the Stone Age, up through the Bronze and Iron Ages, and into the present age of space flight, global communications, genetic engineering, and static-free laundry. It's the mastery of metals that has transformed caves into skyscrapers and clubs into cannons.

Of all the metals, aluminum is the most common in terms of sheer chemical abundance. When mined from its natural habitat, refined, and processed, it's prized for its high strength-to-weight ratio and its ability to resist corrosion. Iron is the second most plentiful metallic element. It's generally considered the most important as well, due to its many applications, especially when mixed with traces of other metals to form various types of steel. Copper is an excellent conductor of heat and electricity and therefore is used in the best cookware and wires. Its unique color also makes it prized in jewelry and other forms of decoration. When exposed to the elements it oxidizes to create an equally attractive patina of greenish malachite. Silver and gold find special uses as coins because of their rarity, in dentistry because of their malleability and durability, and in jewelry because of their brilliance. Platinum is also used in jewelry, but has more practical and important applications as well. It's the platinum in an automobile's catalytic converter than enables the device to transform the most noxious exhaust gases into less dangerous emissions. Like all catalysts, the platinum is needed for the chemical reactions to take place, but is itself not directly involved. The platinum is never used up or chemically altered.

We tend to think of metals as being quite strong and rigid, but this isn't exactly true. In their pure states most metals are actually quite weak. Their molecules, being of identical size, tend to roll and slide over one another because intermolecular linkages aren't very powerful. Alloys, on the other hand, can be formulated to yield amazing properties. By blending molecules of different sizes extra character is imparted to the normally weak metallic crystal. Pure iron is soft. Cast iron, containing about 4 percent carbon, is brittle. When the carbon content is reduced to around 2 percent the product is steel, the tough "skeleton of civilization."

Steel isn't a single substance, but a whole category of alloys, each of which is a mixture of iron and other elements. Within the steel family are titanium steels valued for strength and chromium steels, which are noncorrosive, or stainless. Steels containing nickel are hard, those containing molybdenum are heat resistant, and those containing silicon are flexible. Other types of steel, containing one or more of many other metals, can assume all sorts of specialized properties unknown to any of the individual ingredients alone.

■■■■■■■■■■■■■■■■■■■■■■■■■■■■■■■■■■■■

Gold

The proportion of gold in gold alloys is measured in karats. (Not to be confused with carats, which are a unit of weight used to measure gems. There are 5 carats to a gram or about 140 to an ounce.) Pure gold is 24-karat gold. An alloy containing 75 percent gold is therefore 18-karat gold. Alloys that can legally be called gold range down to 10-karat and are just 40 percent gold by weight.

Gold is further stretched when it's used to cover objects made of lesser metals. Gold-filled objects have a thin layer of gold mechanically brazed to their surfaces and are only about 5 percent gold by weight. Gold-plated objects are coated with as little as seven-millionths of an inch of gold. The gold in so-called gold-washed objects is a mere technicality and quickly rubs off.

Great pains are taken to extend gold because of its high cost, which is a function of both a limited supply and a high demand. Gold is rarer than all but 15 other elements and far from abundant. Its value, however, is due more to its glitter and its status. No one hoards osmium.

■■■■■■■■■■■■■■■■■■■■■■■■■■■■■■■■■■■■

Other metallic alloys contain no iron. Bronze is a mixture of copper and tin and three times harder than either of them. Brass is made from copper and zinc and is more durable than either individually. Solder, an alloy of tin and lead, has a melting point lower than that of either tin or lead alone. Pure gold is very soft. The amount of gold in a man's wristwatch can be pounded into a sheet the size of a double bed or drawn out into a wire three kilometers (two miles) long. To give gold the strength needed to make it useful, it's alloyed with silver and copper to make yellow gold and with copper, nickel, and zinc to make white gold. Silver is similarly alloyed with 7.5 percent copper to form sterling silver, the purest form in which it's generally encountered.

Many other alloys have been developed that have strength, hardness, resiliency, durability, luster, corrosion resistance, melting points, expansion coefficients, densities, malleability, and colors unknown to any naturally occurring metal. Just as the understanding of the polymerization process has yielded a bumper crop of amorphous solids like glass and plastic, so has the understanding of the crystallization process yielded a rich harvest of space-age alloys. By using nature's raw ingredients and elaborating on natural designs we've extensively remodeled much of the material world.

I.21 CAPRICIOUS LIQUIDS

If the solid state of matter is its most dependable, then the liquid state is surely its most capricious. Precariously poised between rigid solids on one hand and ethereal gases on the other, liquids exhibit distinctly schizophrenic tendencies. Like solids, they possess a finite volume. Like gases, they possess no definite shape. Their behavior can best be understood as the interplay of opposing forces, some of which are trying to hold them together and urge them toward solidity, and some of which are trying to rip them apart and vaporize them.

The principal force trying to hold liquids together is called cohesion. This is the same intermolecular affinity that lends solids their firmness. In the case of liquids this force is strong enough to keep neighboring molecules attached to one another and thus provide them with a definite consistency and volume, but weak enough to be partially offset by several competing forces that don't much bother the internal workings of solids.

The first of these counterforces is gravity. A solid stands erect in defiance of gravity because its molecule-to-molecule attractions are stronger than the gravitational attraction of its molecules to the earth. A liquid, however, flattens out and flows under the force of gravity because its internal cohesion is not powerful enough to resist. It's gravity that forces water to the bottom of any container into which it's poured, regardless of its shape. The second force combating cohesion is called adhesion. This is the attraction of the molecules of a liquid to external materials. The water in a paper cup creeps ever so slightly up the sides of the cup due to the adhesion of water to paper.

The delicate ebb and flow of these conflicting forces produce some interesting effects. Most liquids, for example, can act as either adhesives or lubricants. While turning the pages in a book you lick your finger in order to make it sticky. Before you insert the inflation needle into a basketball you lick it in order to make it slippery. In both cases you're completely justified. Spittle's adhesion makes it work like glue while its cohesion makes it work like grease.

Each liquid has its own distinct balance point between adhesive and cohesive urges. Very thick liquids, like molasses, honey, and grease, are dominated by their cohesive tendencies. They lean more toward the solid state than they do toward the gaseous. They flow reluctantly, if at all, resist evaporation, and generally keep to themselves. These liquids are said to be of high viscosity, or very viscous. Some of these liquids are so close to being solids that they have internal forces strong enough to overcome gravity. In these extreme cases the line between the solid and the liquid state is a bit fuzzy.

Low-viscosity liquids, like ether, alcohol, and gasoline, are dominated by adhesive tendencies. They flow readily, penetrate other substances well, and, being closer to gases than solids, evaporate eagerly. Their loosely affiliated molecules easily and frequently wander off into the surrounding air where they are detected as aromas. Adhesive forces, too, are capable of defying gravity. Water is pulled up into sponges, blotters, and treetops by the adhesive forces between itself and the conducting medium. This process, known as capillary action, depends upon the existence of very fine tubes, or capillaries, in which the liquid can hoist itself up in a sort of hand-over-hand climb. Melted candle wax and lamp oil rise in their wicks in this fashion. Towels use capillary action to draw moisture into themselves and away from the object being dried. Cotton towels have lots of little tubes and dry well. Nylon consists of solid rather than hollow filaments and makes a poor towel.

A particular liquid's cohesive/adhesive balance point under any given conditions is sometimes called its wetness. When adhesive tendencies dominate this equilibrium a liquid is more attracted to foreign substances than to itself. A liquid in this situation is said to wet the other substance. Water will wet an unpolished car because the water molecules involved are more attracted to those of the sheet metal than to one another. Water won't wet a polished car. Here cohesive forces dominate. The water molecules prefer their own company to that of the wax. The water withdraws from the wax and gathers itself into little beads, minimizing its contact with the car. By repelling the water, along with all the solvents, corrosives, and abrasives it carries, the wax protects the underlying surface.

I.22 WET, WONDERFUL WATER

Liquids exhibit characteristics of each of matter's other two states. With their finite volumes they resemble solids, with their indefinite shapes they behave somewhat like gases. Resembling each of the other two states, but being neither decisively, liquids can be thought of as matter in a sort of identity crisis. Liquidity, in fact, is really little more than an awkward transition stage between solidity and gaseousness.

The liquid limbo is much avoided by matter. In the universe as a whole, where temperatures tend to be either unbelievably hot or impossibly cold, liquids are virtually unknown. Even here on planet Earth, delicately balanced between the inferno of the sun and frigid expanse of space, liquids are exceedingly rare.

Good old water is pretty much in a class by itself in this regard. Except for some buried petroleum deposits it's the only substance appearing naturally in a liquid state. There are other natural liquids, of course, but they're almost all water-based mixtures, not true substances in a chemical sense. Blood, milk, and other bodily fluids are predominantly water and owe their liquidity completely to their aquatic base. Remove the water and they turn to powder. Most liquids that come from our factories are also basically water, spiked with a few additives. Root beer, soy sauce, ammonia, and vinegar are all water-based solutions, as are latex paints, eyewash, and orange juice. Others, like gasoline, brake fluid, and motor oil, are derivatives of liquid petroleum. A few, cooking oil and turpentine for example, are extracted from organic sources like beans and pine trees. All pale to insignificance when compared to wet, wonderful water in either importance or abundance.

Water, it's commonly understood, is our single most important substance. It's almost synonymous with life as we know it. Earth is populated with everything from protazoa to people because of its water supply. Our search for extraterrestrial life is primarily a search for extraterrestrial water. Life here evolved in the sea, and to this day our bodies are basically just mobile sacks of seawater.

Besides being unequalled in utility, water is also unequalled in abundance. It covers over 70 percent of the globe. Were the earth a smooth sphere instead of a wrinkled mass of mountains and canyons it would be completely covered by a shoreless sea 2,500 meters (8,000 feet) deep. The occurrence of such a critical substance in such a plentiful supply is more than a happy coincidence, it's an essential condition of our being here to appreciate it.

Water's importance, like gold's attraction, is the result of its physical properties. Its exceptionally high capacity for absorbing, storing, and releasing heat moderates the world's climate and makes it more habitable. (See Unit I.15.) Its continual circulation via the rain cycle further equalizes temperatures and humidity. Rain also nourishes and cleanses the world's flora and fauna. Water's ability to dissolve, transport, and deposit nutrients and wastes, on a global scale as well as within a particular living organism, is unmatched. It's the best solvent known. This fact makes pure water a pretty rare commodity. It also makes water extremely vulnerable to pollution. It transports life-threatening and life-giving chemicals with equal facility.

Water is also unique in the way it expands when it solidifies. Virtually every other known liquid shrinks when entering the solid phase. The fact that water grows when it turns to ice, familiar to anyone who has ever stashed a soft drink in the freezer for a quick chill and then forgotten about it, is more than just a curiosity. Freezing water expands with a nearly irresistible force—one that not only shatters pop bottles, but bursts plumbing, destroys roads, pulverizes mountains, and alters landscapes as well.

Effects on the living world are just as great. Because water expands when it freezes, ice floats, and floating ice forms a protective barrier of insulation between the water beneath it and the subfreezing air above. The liquid environment of marine life remains protected from the atmospheric cold. If water behaved like other liquids ice would form at the surface and then sink. Rivers and lakes would freeze solid and many of them would remain that way year-round. Few aquatic creatures could survive such conditions. The ecology of the planet would be dramatically different.

Water, by the uniqueness of both its abundance and its properties, defines the nature of our planet. It's the secret ingredient that bestows the miracle of life on inanimate chemicals. It's the only substance commonly encountered in its solid, liquid, and gaseous states. It's used to calibrate thermometers and as a standard of heat capacity, density, and many other physical constants. Mass and volume standards of the metric system are defined in terms of water. From the structure of its molecule to its ebb and flow in the oceans it's simply the most fascinating stuff there is.

I.23 ELUSIVE GASES

Compared to solids like steel and liquids like water gases hardly seem to be matter at all. Gases have no fixed shape or volume, can't be easily touched or felt, are apparently weightless, and usually invisible. They are, in a word, elusive. Because they couldn't be easily detected, captured, contained, or measured they weren't even recognized as matter by the alchemists. It was the failure to grasp the significance of gases that made combustion so mysterious, allowed the phlogiston theory such popularity, and stymied any progress toward understanding the nature of matter in general. Not until gaseous oxygen and carbon dioxide were discovered by Priestley and Black was Lavoisier able to launch the chemical revolution. (See Unit I.5.)

But despite their ethereal nature gases are true matter in every sense of the word. Gases, just like solids and liquids, have mass and occupy space. Because of their extremely low densities gases usually float above liquids and solids and so appear weightless. But, just like an ocean liner or anything else that floats, gases have a very real and often quite considerable mass. Gases also occupy space even though they lack any definite volume. Gases expand and contract easily to assume a wide range of volumes, but there's a limit to this flexibility. A gas can't be reduced to a volume of zero. No matter how much it's compressed any gas requires some portion of space in order to exist. Under extreme pressures gases simply assume liquid or solid identities and finite volumes.

Gases, like solids and liquids, can be either elements, chemical compounds, or physical mixtures. Gaseous elements are relatively rare. Oxygen, nitrogen, chlorine, fluorine, argon, neon, and a few other less important elements fill this rather small category. The number of compounds that prefer the gaseous state is quite a bit bigger. Carbon dioxide, carbon monoxide, propane, butane, methane, ether, carbon tetrachloride, and a whole host of synthetic substances like Freon are all gaseous compounds. But, because of their restless, uninhibited nature, gaseous elements or compounds, unless collected and carefully stored in an isolated condition, are never found in a pure state. With nothing to prevent them from doing so they all escape into the atmosphere where they diffuse and become part of that great gaseous mixture we call air.

As do the other states of matter, gases exhibit a wide range of properties. Acetylene is violently combustible and burns at such a high temperature that it's used in welding torches. Flourine encourages combustion to such an extent that a jet of it will ignite a piece of wood without a match. Carbon tetrachloride, in contrast, is used in fire extinguishers. Nitrogen and carbon dioxide, among others, are used as propellants for aerosol sprays because they don't react with what they're propelling. Oxygen, on the other hand, is highly reactive. It kills germs, supports life, rusts metals, and helps most fuels burn. Some gases, like nitrous oxide or ether, are used as painkillers or anesthetics by the medical profession. Others, like sulphur dioxide or hydrogen cyanide, are employed against everything from army ants to human armies as lethal weapons.

Despite this wonderful diversity of behaviors and uses, though, all gases are eerily similar in certain other aspects. The reason is that the molecular format of all gases and vapors—like steam, the gaseous state of a substance normally occurring as a solid or a liquid—is virtually identical. Any gas, whether an element, a compound, or a mixture, is a loose collection of individual particles that totally lack any sort of group identity or structure. The molecules of a gas aren't bonded, linked, or even attracted to one another to any meaningful degree. Each acts independently.

The molecules of any gas are not only unfettered, but in rapid and perpetual motion according to the kinetic molecular theory of matter. As they dart about they participate in millions of collisions with their neighbors every second. These intermolecular crashes are perfectly elastic, so no energy is lost in the process. It's a lot like a three-dimensional game of bumper cars played endlessly at a blinding speed. The word *gas,* derived from the Greek *chaos,* is quite appropriate.

As the molecules of any gas charge around they collide not only with one another, but with anything else that lies in their path. The pressure exerted by any gas, whether it be the acetylene inside a steel cylinder or the air inside a birthday balloon, is nothing but the large-scale manifestation of continual and countless molecular impacts. Anything under pressure from a gas is being subjected to the fierce, relentless bombardment of marauding gas molecules, nothing more.

An increase in temperature increases the velocity of molecular movements and the violence of molecular pummeling and pressure. (See Unit II.9.) Heated sufficiently, the molecules of a trapped gas are capable of exerting enough pressure to blow out a tire, rupture a spray can, or destroy a building. Cooling a sample of trapped gas has just the opposite effect. As temperature is lowered so is molecular speed, the force of molecular impact, and pressure. A tire inflated to recommended pressure on a hot summer day will be underinflated the following winter even though not a single molecule of air has escaped. The lower winter temperature simply reduces the pressure of the enclosed air.

When gases aren't confined in rigid containers an increase in temperature will cause them to expand. Again, added heat gives extra zip to the molecules. As the molecules fly about at ever greater speeds the range of their motion increases and they occupy more space. When the air in a hot-air balloon is heated it expands. It becomes less dense than the surrounding unheated air and floats atop it like a buoy on a lake, lifting the gondola and its contents along with it.

• •

Molecular Math

Boyle's law relates the volume of a gas to the pressure under which it's kept. Charles's law relates the volume of a gas to the temperature at which it's kept. Together they're known as the universal gas laws and state that the extreme elasticity of gases make any measurement of their volume meaningless unless the conditions under which it is taken are specified.

Amedeo Avogadro (1776–1856), an Italian chemistry professor, was working in strict accordance with these laws when he added one volume of nitrogen to two volumes of hydrogen. The gases failed to react chemically, and he ended up with a mixture that measured three volumes. No surprise. He then combined one volume of nitrogen with two volumes of oxygen. These reacted to form a new gas, nitrogen dioxide. Interestingly enough, this new substance measured two volumes, not three.

Avogadro correctly deduced that the volume of a gas depends upon the number of molecules it contains. When nitrogen and hydrogen mix there is no chemical reaction and no change in either the number of molecules or volumes involved. But, when nitrogen and oxygen react chemically the volume is reduced from three to two as a result of a corresponding reduction in the number of molecules. More specifically, one molecule of nitrogen reacts with two molecules of oxygen to form two molecules of nitrogen dioxide.

According to the theory of matter in Avogadro's day, though, this was impossible. It was assumed then that all elements, including nitrogen, existed as unattached atoms. So how could a single atom of nitrogen be used to create two molecules of nitrogen dioxide?

Avogadro was the first to realize that nitrogen, oxygen, and hydrogen commonly exist not as single atoms but as molecules consisting of two identical atoms, and thereby provided Mendeleev with the proper atomic weights he needed to construct his chart.

—What happens to a given volume of common oxygen (O_2) when it is changed to ozone (O_3) by solar radiation?

—The atomic weight of oxygen is 16 atomic mass units (amu). What weight do you suppose was assigned to oxygen before Avogadro's work? What is the weight of a molecule of ozone?

—Can you write a balanced chemical equation for the formation of ozone from common oxygen? What about the formation of nitrogen dioxide? Water?

• •

The heat that powers the expansion of a sample of gas is unneeded and unwanted when the gas contracts. As any gas is forced into a smaller volume it releases this excess heat by giving it up to its immediate environment. A bicycle pump gets hot when you inflate a tire because the air passing through it is being compressed. A sample of gas allowed to expand on its own, on the other hand, experiences a drop in temperature as its molecules spread out, collide less often, and slow down. The spray coming from an aerosol can not only feels cool, it is cool. The escaping gas cools as it expands.

I.24 AN OCEAN OF AIR

Air holds the same unique status among gases that water does among liquids— it's at once the most abundant and dominant member of its class. Like water, air is essential to our survival as individuals and a major shaper of the environment in general.

Although colorless, odorless, tasteless, silent, and seemingly weightless, air was quickly recognized as a real substance by the Greeks. Anaximenes, you may recall, went so far as to postulate that air was not only something, but everything. Empedocles included air as one of his four elements after pointing out that air trapped underwater in an inverted cup resisted the intrusion of water and therefore had volume and truly was some sort of stuff. Aristotle affirmed the existence of air when he sanctioned the four elements of Empedocles. Aristotle also announced that air was hot and dry, and a weightless balance of gravity and levity.

Aristotle's definition of air survived pretty much intact until it fell under the scrutiny of Galileo. In this case Galileo's interest was aided and his efforts abetted by one of his prize pupils, Evangelista Torricelli (1608–1647). After some preliminary investigations Torricelli hypothesized that, despite Aristotle's assertions, air actually did have weight, and that it was this weight that made drinking water through a straw possible.

To test this new idea Galileo simply took an apparently empty flask and weighed it. He then forced additional air into the flask with the aid of a pump, sealed it, and weighed it again. The flask, he discovered, weighed more when crammed full of extra air than it did when it was filled with nothing but its normal atmospheric quota. The difference in the two weighings was that of the added air. No one else had ever noticed air's weight, he explained, because it was equally present everywhere.

The fact that air had a real and definite weight had major implications. To begin with, it suggested that the atmosphere had real and definite dimensions. Air didn't crush everything beneath it. It's weight was fairly limited, and so must be the extent of the atmosphere. Air was a somewhat local phenomenon, not the universal stuff of the heavens as many had supposed. "We live submerged at the bottom of an ocean of air," is how Torricelli put it.

This invisible atmospheric ocean is held to the earth by the force of gravity

■■■■■■■■■■■■■■■■■■■■■■■■■■■■■■■■■■■■■

The Atmosphere

Although it's a seamless sea of air, the atmosphere is generally broken down into a number of layers for academic purposes. In reality each of these layers is very indistinct and merges imperceptibly with its neighbors.

The lowest layer is called the troposphere. This is where there's enough oxygen to support life and where weather happens. Above that lies the stratosphere, where temperatures remain at subarctic levels. This is where airliners fly, both to avoid varying weather conditions and to ride on swift jet streams. The upper atmosphere consists of the mesosphere, the ionosphere, and the exosphere. These outer layers protect the earth from the cosmic rays of outer space, the ultraviolet radiation of the sun, and errant meteors.

The atmosphere as a whole acts as a giant blanket that insulates the planet from the bleakness of outer space and helps it retain the radiant warmth it receives from the sun. At higher altitudes this blanket thins out, making mountaintops cold enough to retain year-round snow, even on the equator.

■■■■■■■■■■■■■■■■■■■■■■■■■■■■■■■■■■■■■

in much the same way as is its more tangible aquatic counterpart, water. It's the air's attraction to the earth via the force of gravity that accounts for its weight. And, as Torricelli hypothesized, it's this weight, this atmospheric or barometric pressure, that properly explains the operation of a drinking straw.

Coca-Cola is held in a glass not only by gravity, but by a weighty blanket of air. Sucking on a straw removes some of the air from inside it and reduces this downward force. The Coke outside the straw, subject to full atmospheric pressure, outweighs the Coke inside the straw by the amount of air removed. The heavier, outside Coke pushes the lighter, inside Coke up the straw in much the same way that a big brother at one end of a teeter-totter pushes up his little sister at the other.

Air, like any other fluid, flows immediately and with considerable force under the influence of gravity. Just as water flows in an attempt to equalize its level, so air flows in an attempt to equalize pressure. On a small scale this allows us to breathe. We don't really suck in air, rather it's barometric pressure that literally rams air down our throats when we expand our lungs to create a low-pressure area.

On a large scale this same mechanism creates a good deal of our weather. Air flowing from high- to low-pressure areas is the source of all our wind as well as most of our precipitation. Low-pressure areas originate when the density of air is reduced. This happens in several ways, the most common being the warming of the air by sunshine. (See Unit I.23.) Additionally, because the density of water vapor is less than that of air, an increase in humidity will also reduce the overall density of an air mass. Hot, wet air, being relatively light, constitutes a low-pressure system. Cool, dry air, being relatively heavy, constitutes a high-pressure system. Air, in an effort to keep pressure everywhere uniform, rushes from high-pressure areas to low-pressure areas as wind. As the various air masses collide the hot, wet air is cooled and its ability to retain moisture is diminished. The moisture it can no longer hold falls as rain or snow. (See Unit II.14.)

Besides heat and humidity there's a third factor that reduces air pressure: velocity. A body of air set in motion has a lower pressure than it did standing still. The higher the velocity, the lower the pressure. This interesting phenomenon was discovered by a Swiss scientist named Daniel Bernoulli (1700–1782). Seemingly trivial, the Bernoulli effect allows planes to fly.

The bottom surface of a plane's wing is essentially flat. Its top surface, in contrast, bulges slightly upward. As a plane moves forward, then, the air immediately beneath its wings remains undisturbed while the air immediately above its wings is pushed up and out of the way. The lower air remains still while the upper air moves. The movement of the upper air lowers its density according to the Bernoulli effect. The denser air beneath the wing tries to equalize this inequality by flowing upward. As it does so it lifts the wing and the attached plane along with it. A plane can't hover because once the forward motion of the wing ceases so does the inequality of air densities.

The rotor on a helicopter also works on the Bernoulli principle. It is, in effect, a cluster of tiny wings set into motion by a motor of some sort. In this case it's the circular motion of the rotor blades rather than the forward motion of the plane that puts the forces generated by the Bernoulli principle to work. A helicopter can hover because even as it does so its "wings" are still swirling through the air generating low-pressure areas above them and therefore being pushed upward by the relatively high-pressure areas beneath them. Should the rotor stop spinning the helicopter will crash just as surely as a plane will when it loses its forward velocity.

Wind also provides an increase in air speed and a resultant drop in pressure in accordance with the Bernoulli principle. Strong winds have been known to "suck" the roofs off buildings. Technically, of course, it's not the pull of the outside wind but the push of the air from within the structure that's responsible. The roof is actually blown off by air that is trapped inside the house and trying to get out. Tornado warnings usually include instructions to open all windows in an effort to help air pressure equalize less violently.

I.25 TEMPORARY COMPROMISES

Despite the apparent orderliness of a gaseous atmosphere wrapped around a liquid hydrosphere and a solid lithosphere, things aren't quite that simple.

They never are. The reason is that there are always many conflicting forces and opposing processes taking place simultaneously in the physical world. The result is that things and events exist not in pure forms, but in states of equilibrium that are temporary compromises among the competing pushes and pulls. And so, just as the elements have a penchant for uniting chemically to form compounds—and compounds in turn crave to mingle physically to form mixtures—gases, liquids, and solids love to cross state lines.

There are two basic types of interstate combinations: solutions and colloids. Solutions consist of a primary substance called a solvent and a secondary substance called a solute. The solvent and the solute come together each in its own state, but the resulting solution takes on the characteristics of the dominant solvent. Saltwater is a solution of a liquid solvent—water—and a solid solute—salt—and behaves entirely as a liquid.

Colloids consist of a primary substance called the dispersing phase and a secondary substance called the dispersed phase. The two phases again come together each in its own state, but in this case the resulting colloid may either assume the state of the dispersing phase or an entirely different physical condition that may be a bit unconventional if not downright weird. A marshmallow is a colloid of gaseous air dispersed in solid sugar, but is itself neither a solid nor a gas. It's just a colloid that defies any attempt at categorization.

The main difference between solutions and colloids in a technical sense is the size of the particles dissolved or dispersed. In a solution the solute is completely disassembled into its individual and invisible molecules. The dispersed phase of a colloid, on the other hand, exists as tiny clumps of molecules that are of a size that usually puts them right on the threshold of visibility. Both the dissolved particles in a solution and the dispersed particles in a colloid are small and light enough so that they float among the particles of the solvent or dispersing phase, respectively. For this reason neither solutions nor colloids will separate into their component substances merely by standing. Their secondary substances will never settle out via what's called sedimentation. Anything labeled "Shake Well Before Using" separates under the influence of gravity and isn't a true interstate combination. Pepto-Bismol, salad dressing, and spray paint are neither solutions nor colloids. They're just collections of substances in different states coexisting in a common container.

Solutions come in many forms, but are most common with liquid solvents, especially water. Water, like other less common solvents, is capable of dissolving gases, other liquids, and solids. Household ammonia and chlorine bleach are solutions of gaseous ammonia or chlorine, respectively, in liquid water. Soft drinks, too, are solutions of a gaseous solute, this time carbon dioxide, in a liquid water solvent. Even before a trip to a bottling plant, though, water carries dissolved portions of the air. Fish extract dissolved gaseous oxygen from water when they breathe.

Liquids are a bit particular about forming solutions with other liquids. Liquids that will dissolve one another are said to be miscible, others immiscible.

■■■■■■■■■■■■■■■■■■■■■■■■■■■■■■■■■■■■

Soap

Soap molecules have hydrophobic "heads" that hate water and hydrophilic "tails" that love it. When a soap and water solution is applied to a dirty shirt, the hydrophobic "heads" of the soap molecules burrow as far as they can into the shirt in order to escape the water. As they do so they break the dirt into tiny specks, which they surround like a school of feeding piranhas. The hydrophilic "tails" remain attracted to the water and keep the dirt particles suspended so that they can be carried away in the rinse.

Detergents are just synthetic soaps and work the same way. They're generally more effective than natural soaps, especially in hard water where soap molecules tend to attack the minerals in the water rather than the targeted dirt. When soap is used in hard water the result is something called soap curd or scum, the cause of bathtub rings.

Suds aren't necessary for either soap or detergents to clean, but are so psychologically important that foaming agents are often added to sudsless cleaners in order to make them more satisfying.

■■■■■■■■■■■■■■■■■■■■■■■■■■■■■■■■■■■■

Water and acetic acid are miscible and form vinegar when mixed. Vinegar is a true solution that won't separate while standing on your cupboard shelf. Salad oil and water are immiscible. They can be mixed, but will separate again spontaneously.

The solution of solids in liquids also occurs rather selectively. (See Unit I.16.) When a solid is soluble in a liquid it generally dissolves more readily at elevated temperatures. It's easier to dissolve a spoonful of sugar in hot tea than in iced tea. A liquid solvent, regardless of its temperature, though, will dissolve only so much of a solid solute and no more. When it has dissolved all the solute it can it's said to be saturated. Solute added to a saturated liquid solvent won't dissolve. Extra sugar added to a saturated cup of tea, whether hot or iced, will simply remain as solid sugar and settle on the cup's bottom.

Colloids, like solutions, most often have a liquid as their primary substance, or dispersing phase. The dispersed phase, again, can exist in any state. Most foams, like shaving cream or whipped cream, are colloids of a gas, usually nitrogen or carbon dioxide, dispersed in a liquid. Milk, coffee, and muddy water are all colloids with a solid phase dispersed in water. Liquid-liquid colloids are also common, but often require the presence of a third party, called an emulsifier. Salad oil and water by themselves are immiscible, but when emulsified by an egg yolk form colloidal mayonnaise. Soaps and detergents are important emulsifiers. With their aid water can disperse and wash away otherwise impenetrable dirt and oil.

Colloids with gaseous and solid dispersing phases are common, too. Smoke and dust are solid colloidal particles suspended in gaseous air. Fog and clouds are colloids of liquid water droplets similarly suspended. Bread, like marshmallow, is a gas-in-solid colloid, while butter is a liquid-in-solid version. Colored plastics consist of solid pigment particles combined colloidally with the solid plastic.

I.26 SUBATOMIC FRAGMENTS

As the nineteenth century wound down, so did much of the excitement about exploring matter. Lavoisier had discovered the elements and Dalton had discovered the atomic mechanism by which they interacted. The gaps in Mendeleev's chart were being inexorably filled. Matter was being neglected in favor of more glamorous subjects. It was while investigating electricity, very glamorous and still something of a novelty in the 1890s, that Joseph J. Thomson (1856–1940), an Englishman, make a discovery that would thrust matter once again into the scientific spotlight.

During the course of many electrical experiments Thomson passed an electric current lengthwise through a sample of gas sealed in a hot dog–shaped glass tube. When he did no he noticed a definite *something* flowing from one end of the tube to the other—a something that bore no resemblance at all to any solid, liquid, gas, solution, or colloid he had ever encountered. Thomson was able to determine that the something was a stream of tiny particles and that each of the particles carried a negative charge. He also succeeded in computing the mass of one of the mysterious particles. Much to his amazement it was only a small fraction of the mass of a hydrogen atom, which up until that moment was assumed to be the lightest thing in existence. Thomson repeated his experiment with a number of different gases and each time the result was the same. The little particles always appeared, always carried a negative charge, and always weighed the same minuscule amount. He had no idea what he had discovered, but named the particles electrons after their electrical source.

Meanwhile, in France, Antoine-Henri Becquerel (1852–1908) was conducting various experiments with equally if not more glamorous x-rays, recently discovered quite by accident. (See Unit II.25.) During the course of his investigations Becquerel had a little accident of his own. He left some undeveloped photographic plates inside his desk drawer along with some salts containing uranium. When he later retrieved these materials he was surprised to find that the film had somehow been exposed. He didn't know what to make of the strange event but his assistant, Marie Sklodowska Curie (1867–1934), theorized that the uranium had, without any sort of prompting, thrown off radiation of some sort and that the uranium-generated radiation had exposed the film in much the same way that ordinary light would have.

Sensing the significance of this development, Curie, along with her husband, Pierre (1859–1906), dedicated herself to finding other substances

■■■■■■■■■■■■■■■■■■■■■■■■■■■■■■■■■■

The Curies

Marie Sklodowska was born the daughter of teachers in Warsaw in 1867. Pierre Curie was born the son of a physician in Paris in 1859. They met while studying chemistry and physics together in Paris and were married in 1895.

Together with their mentor, Antoine-Henri Becquerel, they received the Nobel Prize in physics in 1903 for their discovery of radium and polonium. Eight years later Marie Curie won an unprecedented second Nobel Prize, this time in chemistry, for her continued investigations of the same two elements.

The Curies had a daughter, Irène, born in 1897. She married Frédéric Joliot, and together the Joliot-Curies also won a Nobel Prize in chemistry in 1935 for their work with artificial isotopes.

Because the dangers of radiation were at first unknown, many of its early investigators suffered radiation poisoning and early deaths. Becquerel died at the age of 56, Pierre Curie at 47, and Frédéric Joliot-Curie at 58. Irène Joliot-Curie died at 59 of leukemia, the same disease that claimed her mother at age 67.

■■■■■■■■■■■■■■■■■■■■■■■■■■■■■■■■■■

with this built-in capacity to generate radiation, what we now call radioactive substances. A careful scrutiny of all known elements yielded only thorium. Four years of intense research and demanding physical labor resulted in the discovery of two new elements—polonium, which she named for her native Poland, and radium—each of which emitted radiation similar to that thrown off by uranium, but hundreds of times more powerful.

One of Thomson's students, a New Zealander named Ernest Rutherford (1871–1937), then incorporated the Curies' findings into his mentor's experiments by zapping gases not with electricity but with the radiation emitted by uranium and thorium. Using this procedure he identified two distinct types of radiation coming from these substances that he labeled alpha and beta after the first two letters of the Greek alphabet. Alpha radiation, he found, consisted of relatively slow-moving, heavy particles and was positively charged. Beta radiation consisted of faster, lighter particles and was negatively charged. Beta radiation, in fact, was the same thing that Thomson had discovered earlier. It was electrons. Rutherford named the heavy and equally but oppositely charged particles of alpha radiation protons.

Further investigations soon revealed that electrons and protons were actually subatomic fragments. The almighty atom, contrary to the creed of Democritus, Dalton, and everyone up to that time, wasn't the ultimately small particle of matter. Atoms were divisible into at least two component pieces.

Nor were atoms indestructible. The source of alpha and beta radiation was the direct result of atoms that fell apart as they underwent radioactive decay. Nor was the atom immutable. Atoms of uranium, polonium, and thorium, once they threw off radiation and disintegrated, turned into atoms of other, lighter elements. The atom—long assumed to be indivisible, eternal, and immutable—was suddenly vulnerable, temporary, and unstable.

I.27 A MINIATURE SOLAR SYSTEM

The air of complacency about matter was suddenly whipped up into a storm of excitement. Proposing models of the atom and trying to figure out how it worked became the most glamorous of all scientific endeavors. The earliest attempts relied more on fertile imaginations than assembled facts, but the accumulation of knowledge and the subsequent refinement of atomic conceptualizations proceeded rapidly.

Thomson's electrons were light and negatively charged; Rutherford's protons were heavy and positively charged. Their charges were equal but opposite. Atoms contained both electrons and protons but were themselves neutral. They therefore contained equal numbers of electrons and protons whose opposing charges offset one another. That's about all that was known for sure about atomic structure at first.

Then Rutherford bombarded a thin piece of gold foil with some protons in the form of alpha radiation. Out of every million protons fired at the foil all but 100 sailed through cleanly without so much as slowing down. Of the remaining 100, 99 were deflected from their original paths as they passed through the foil while the hundredth, or one out of the original million, was bounced back like a tennis ball off a brick wall. Using logic and mathematics, Rutherford deduced that the gold atoms in the foil weren't solid chunks of matter, but airy constructions that consisted mostly of empty space. Almost the entire mass of the atom seemed to be concentrated in a tiny, incredibly dense central core. The core, being massive, was obviously the site of the protons. It was this small solid core, or nucleus, that deflected similarly charged protons passing nearby and repelled the occasional direct hit. The light, negatively charged electrons evidently occupied the outlying terrain of the atom. Protons passing through this area were unaffected.

Starting with this general layout, a Dane, Niels Bohr (1885–1962), ingeniously determined that the electrons continuously circled the nucleus and did so in specific orbits, which corresponded to various energy levels. It had long been known that each element, when heated, emitted its own unique pattern of light waves, or spectrum. (See Unit I.8.) Conversely, a cool sample of an element absorbed the exact same pattern of light waves that a hot sample radiated. Bohr deduced that the emission of light resulted from an electron giving up energy as it fell from a distant orbit into lower one. An electron jumped back into a higher orbit by absorbing an identical amount and type of light.

By using the known mass of electrons and the wavelengths of the light involved in the change of orbits, Bohr was able to construct a model of the atom that resembled a miniature solar system. The atom of each element turned out to have electron orbits that were distinctive in size, number, and arrangement and that perfectly explained the unique spectrum associated with each element. The arrangement of electrons also effectively explained the sizes and numbers of the rows and columns of Mendeleev's periodic chart. A periodically repeating pattern of electron configurations was responsible for the periodically repeating occurrence of chemical properties.

It was now obvious that it was the interactions of electrons from different atoms that constituted chemical reactions. It was the outer electron clouds that bonded, merged, parted, repelled, and attracted as atoms underwent chemical changes, became molecules, and formed compounds. It was electron clouds that linked up in geometric grids to form crystals. The nucleus, buried deep inside the atom, didn't seem to participate in chemical events or be much affected by them.

Back in Rutherford's lab Henry Gwyn Jeffreys Moseley (1887–1915) was analyzing the spectra of atoms with x-rays rather than with visible light. Based on his findings, the 26-year-old physicist formulated an equation that related an atom's x-ray spectrum to the number of protons in its nucleus. Further investigations yielded startling results: hydrogen, the first element in the periodic table, had one proton; helium, the second element, had two protons; lithium, the third element, had three protons, and on and on up through the chart. There was an exact, unerring correlation between an element's ordinal position in the chart and the number of protons in its nucleus.

The elements now each had a precise and unalterable proton count and could be lined up indisputably and irrevocably from number 1, hydrogen, to number 92, uranium. It was proton count, or atomic number, not atomic weight, that was the final determinant of elemental order and identity. The escalation of atomic weight was only a secondary coincidence that, as Mendeleev had long ago discovered, didn't always hold true. Without knowing exactly why, he had properly placed tellurium, with its 52 protons, before iodine, with 53, even through the former slightly outweighed the latter.

But what, then, was the origin and meaning of the atomic weights used since the time of Dalton to identify and organize the elements? Atoms aren't weighed in ounces or grams, of course, but in something called atomic mass units, or amu for short. Hydrogen, the lightest element, has a mass of 1 amu. Hydrogen, element number 1, has one proton in its nucleus and one electron orbiting it. The weight of a proton, roughly speaking, is 1 amu. An electron weighs virtually nothing. The single proton therefore accounts for all the mass of a hydrogen atom and a hydrogen atom can be assumed to consist of one proton, one electron, and little else. The arithmetic works. Element number 2, helium, has two protons and two electrons. If it, like hydrogen, were constructed out of nothing but protons and electrons its atomic weight should be 2 amu. But it isn't. Helium has an atomic weight of 4 amu. There are 2 amu of mass in a helium atom that are unaccounted for by protons and electrons. Every element except hydrogen, in fact, has an atomic weight that exceeds the combined weight of its protons and electrons. They all contain something else besides these two components.

In 1932 a British physicist, James Chadwick (1891–1974), found this something else. He called it the neutron because it was electrically neutral. The neutron, like the proton, resides in the atomic nucleus. The neutron, like the proton, has a weight of approximately 1 amu. A helium atom has two protons and two neutrons and weighs 4 amu. The atomic number of an element denotes its proton count; its atomic weight denotes its proton-plus-neutron count.

The discovery of the neutron solved two other atomic riddles besides atomic weight. First, it had never been clear what held an atom's protons clumped together in the nucleus. Like charges normally repel one another, and here were all these similarly charged protons packed shoulder to shoulder. The neutron, it seems, is the glue within the nucleus.

Secondly, the neutron explains the fact that not all atoms of an element have the same weight. This curiosity was first noted by Thomson, who found some chlorine atoms with a weight of 37 amu and others with a weight of 35 amu. Both were chlorine atoms; both had 17 protons and 17 electrons. Both bonded with sodium to form salt. The only difference between them, now understood, was that one contained 20 neutrons and the other only 18. A new word, *isotopes,* was coined to describe atoms that had identical numbers of protons and were thus chemically identical, but had different numbers of neutrons and different weights.

I.28 NUCLEAR PHYSICS

As might be expected, the knowledge we've accumulated about how the atom is put together and operates has given us an enormous amount of power over the material world. We are, in fact, now capable of determining the age of fossils, attaching a retina with a beam of light, following invisible atoms as they travel through the human body or around the world, transforming one element into another, creating brand-new elements, and blowing ourselves and our entire planet to bits.

The nature of all these wonders places them beyond the bounds of traditional chemistry and within the area now commonly called nuclear, or subatomic, physics. Whereas chemistry deals with whole atoms as they participate in chemical events like combustion and physical events like melting, nuclear physics deals with events that take place wholly within individual atoms.

Some of our subatomic accomplishments consist of merely properly understanding and meaningfully interpreting natural events. Carbon, we've come to realize, exists in two isotopic forms: one with an atomic weight of 12 amu (carbon-12) and the other with an atomic weight of 14 amu (carbon-14). Atoms of carbon-14 are created by cosmic radiation in the upper layers of the atmosphere and are exceedingly rare, existing only to the extent of about one out of every trillion atoms of carbon. These freak carbon atoms circulate through and become a part of living creatures via the respiration and photosynthesis processes. The carbon in a living organism, which is a product of its environment, is one-trillionth carbon-14.

Besides being rare, carbon-14 atoms are also unstable. They radioactively decay into atoms of carbon-12. After an organism dies, then, its intake of carbon-14 atoms ceases and those it already contains start to decay into the more ordinary carbon-12 variety. Its ratio of carbon-14 to carbon-12 starts to diminish. Because we know the rate at which carbon-14 degenerates into carbon-12, the time elapsed since an organism's death can be determined by measuring the ratio of its two carbon isotopes. The so-called carbon-14 dating process, developed at the University of Chicago in 1947, has been an invaluable tool for archeologists, anthropologists, and paleontologists in determining the ages of excavated scraps of bone, wood, fibers, and other once-living materials.

Other wonders of nuclear physics are the result of the artful guidance of natural processes in new directions. Electrons routinely shuttle back and forth among various orbits within the atom. Under normal conditions they do so as an uncoordinated collection of individuals, randomly absorbing bits of energy as they climb to a higher orbit and releasing energy again as they fall to a lower one. It's this behavior, first explained by Bohr, that accounts for the identifying spectra of the elements. The same behavior when orchestrated by nuclear physicists turns light into a magic wand.

By synchronizing electrons so that they all fall in unison, all the bits of energy they release reinforce one another and produce light beams of incredible strength. These beams are called lasers, the word *laser* being an acronym for Light Amplification by Stimulated Emission of Radiation. Lasers are now commonly encountered in grocery stores where they ring up prices, in homes where they play back music recorded on compact discs, in hospitals where they've replaced the scalpel, in dentist's offices where they've replaced the drill, and in countless industrial, commercial, and technical operations.

The radioactive decay of uranium is another example of a naturally occurring subatomic process that has been tamed and exploited. The same energy that fogged the film in Becquerel's drawer, greatly concentrated, is now used to power submarines, generate electricity, and threaten us with extermination. (See Unit II.31.)

Perhaps the most impressive nuclear achievements are those that are entirely the work of human beings. The most useful of these involve the

● ●

Nuclear Medicine

Although it is hardly a pleasant thought, imagine that you feel constantly on edge—fidgety and anxious. You're having trouble sleeping and, even though you're eating normally, you keep losing weight.

Your doctor suspects you have a hyperactive thyroid. This gland, located at the base of your neck, controls your rate of metabolism by means of a hormone it produces called thyroxine. Generally speaking, the more thyroxine your thyroid secretes the faster your body works. Yours seems to be going too fast.

In order to confirm her diagnosis, the doctor instructs you to drink a salty liquid containing a tracer dose of iodine-131, a radioactive isotope produced especially for medical use. The thyroid needs iodine to manufacture thyroxine, and the amount of I-131 it absorbs will indicate its activity level.

The next day you report to a hospital where, using a Geiger counter, the amount of I-131 in your thyroid is measured. As suspected, your thyroid has absorbed an unusually large amount of iodine during the previous 24 hours. It's working overtime.

The doctor now prescribes a series of stronger, therapy doses of I-131. Like the tracer dose, these are absorbed by the faulty gland. The radioactive iodine proceeds to decay, throwing off a number of sub-atomic particles and powerful gamma rays as it does so. This radiation actually destroys a portion of your thyroid, thereby crippling it and reducing its output of thyroxine. Soon you are yourself again. Thanks to nuclear medicine, you've been able to avoid surgery.

Iodine naturally exists only as I-127. This stable isotope is transformed into I-131 by exposing it to a nuclear reaction such as that which takes place inside a nuclear generating plant. The resultant I-131 has a relatively short half-life of eight days. That is, one-half of its atoms radioactively decay every eight days. (See Unit IV.4.)

—What are the advantages and disadvantages of a half-life of eight days?

—Would I-131 be more or less effective for treating a hyperactive thyroid if it had a half-life of two days? Two weeks?

—Can I-131 be used to diagnose hypothyroidism, where the thyroid works too slowly? Can it be used to treat such a condition?

● ●

creation of artificial isotopes. It might seem that there are already plenty of isotopes without adding to the confusion. The 92 natural elements exist in something like 350 isotopic forms. Nearly every element has at least a couple of isotopes and some, like uranium with its 14, have considerably more. The big difference between natural and artificial isotopes, though, is that the former are, with a few important exceptions like uranium, stable while the latter are, for the most part, unstable.

When loaded up with extra neutrons even common substances like oxygen can be made radioactive. And radioactive oxygen, just like uranium or radium, decays and emits radiation as it does so—radiation that can be detected and measured. Substances tagged with a population of unstable, or radioactive, atoms can be traced as they participate in chemical reactions, circulate through the body, or travel through the environment. The ability to tag and follow individual atoms has generated enormous amounts of knowledge about the way things work.

We also have the ability to add extra protons to an atomic nucleus. By this means it's theoretically possible now to fulfill the alchemists' dream and turn lead into gold. But gold built an atom at a time costs far more than it's worth, and the transformation of one element into another is of little practical value.

By adding protons to uranium atoms we're capable of creating entirely new elements with atomic numbers higher than 92. The production of these so-called transuranic elements, with names like curium, einsteinium, and mendelevium, is an academic exercise for the most part. One artificial element, though, is of utmost importance. Plutonium, a by-product of some uranium-based nuclear reactions, has many potentially profound applications. (See Unit II.31.)

As the atom is subjected to more meticulous scrutiny, split into ever-smaller fragments, and further fiddled with, it will no doubt yield many more marvels. Nuclear physicists have already identified over 100 subatomic particles, many of which defy the laws of traditional physics. But the exploration of the atom is driven by a search for simplicity, not complexity. It's driven by the need to understand matter and reduce it to knowable terms. There's an almost inviolate assumption that somewhere deep inside the atom resides a comprehensible answer to the age-old question: What is stuff?

II

ENERGY

"All is flux, nothing stays still."
—Heraclitus, c.540–c.480 B.C.

II.1 THE AGENT OF CHANGE

The physical world most readily makes itself apparent as clouds, pools, and chunks of matter. Of that there can be little doubt. Elements, compounds, and mixtures existing as gases, liquids, and solids are the stuff of which our atmosphere, our oceans, and the earth beneath our feet are made. The physical world, at first glance anyway, would seem to be a material world and nothing more.

But air, water, and dirt don't just sit there. The wind blows, rivers flow, and the earth quakes and erupts. A cloud explodes in a burst of lightning, rain falls, and a forest burns. Even rocks eventually crumble to dust and are washed to the sea. Everything, it seems, is either coming or going, ebbing or flowing, shrinking or growing. Matter, regardless of its identity, format, or location, is forever being subject to change.

The agent of that change is energy. While matter accounts for what things are made out of, energy accounts for what happens to them and what they do.

Matter and energy—what stuff is and what it does—are now commonly considered to be two quite different aspects of the physical world. The line dividing them, though, is a rather fuzzy one and at times difficult to draw. The discussion, definition, and even existence of one inevitably involves the other. Like form and substance or speed and direction, they're intimately related. Until fairly recently, in fact, they were completely, and quite understandably, confused.

The first humans, of course, made no attempt whatsoever to distinguish matter from energy. The external world in those days was far beyond comprehension and completely exempt from any sort of analysis. But even the Greeks, with all their philosophical probings, failed to differentiate energy from the matter through which and upon which it acted. Empedocles included fire, the most spectacular form of energy, with air, water, and earth as one of the four basic substances out of which the entire material world was constructed. Aristotle thought that heat, the most common form of energy, along with cold, humidity, and dryness, was one of matter's four primary properties. Invisible, intangible energy was never thought of as anything apart from more obvious and concrete matter, but always as something encompassed within it.

The alchemists, Galileo, Boyle, and Dalton also failed to discriminate between energy and matter. Even the brilliant Lavoisier, perhaps because of his strong chemical mind-set, listed among his elements something called caloric—an invisible, weightless, and surprisingly phlogiston-like fluid that supposedly flowed into and then out of substances making them hot or cold as it did so. Like his other elements, caloric was considered to be immutable, indestructible, and completely material in nature. There was a finite amount of caloric in the world, Lavoisier postulated, and all the heatings and coolings that took place were merely redistributions of the same basic supply.

■■■■■■■■■■■■■■■■■■■■■■■■■■■■■■■■■■■■

Benjamin Thompson

Thompson was originally an American, born in Woburn, Massachusetts. A stout Tory, he fled the New World for London when the British evacuated Boston in 1776. From England he made his way to Austria where, at the request of Prince Maximilian, he entered the Bavarian civil service. He spent 11 years there serving in various capacities and performing a number of public services.

Because of his accomplishments in both Austria and England he achieved a variety of distinctions. He was knighted by King George III to become Sir Benjamin Thompson and then later was made a count of the Holy Roman Empire, whereupon he began calling himself Count Rumford after his onetime hometown Rumford (now Concord), New Hampshire. He also spent some time in France where he met and then married the widow of the beheaded Lavoisier. Thompson lies buried just outside of Paris.

More than for his lifestyle, however, Thompson (or Rumford) is remembered for his 1798 presentation to the Royal Society of London, "An Experimental Enquiry Concerning the Source of the Heat which is Excited by Friction."

■■■■■■■■■■■■■■■■■■■■■■■■■■■■■■■■■■■■

It wasn't until nearly the end of the eighteenth century that gadabout physicist-philanthropist Benjamin Thompson (1753–1814) finally suggested that heat, or caloric, might not be a material substance. While supervising the boring of cannons for the Bavarian army he noted that heat was produced during the process in apparently limitless quantities. Examination of the brass before and after the boring revealed that no material change had taken place. No brass had been consumed and there had been no chemical reaction. The heat, it seemed, sprang out of nothing but the act of cannon-boring. It didn't flow in from neighboring materials, because nothing nearby got cooler as the cannons got hot. Heat, caloric, or whatever, wasn't being redistributed, but created right there before his eyes. If it wasn't the product of chemistry and wasn't being borrowed from anything else, where was all this heat, or caloric, coming from?

Thompson correctly hypothesized that the heat was the consequence of nothing but the mechanical motion involved in boring the cannons. The motion became heat while the brass remained brass. There was a physical change of some sort taking place, but it didn't directly involve matter. It involved the transformation of one form of energy into another. Because energy was able to participate in a nonmaterial physical event, Thompson claimed, it was a physical commodity in and of itself and something apart from matter.

II.2 THE CONSERVATION OF ENERGY

Once commenced, the disentanglement of energy from matter proceeded rapidly. Sir Humphry Davy (1778–1829), an English gentleman-chemist credited with the discovery of no fewer than 12 elements, substantiated Thompson's observations by turning motion into heat under laboratory conditions. All he did was rub a couple of ice cubes together and watch them melt. The motion of the rubbing was transformed into the heat needed to melt the ice. The water, although warmed, remained unaltered in either amount or chemical composition. Motion and heat affected the material world as external forces, but weren't really included within it. Energy wasn't matter. Caloric, just as phlogiston had been earlier, was struck from lists of the elements.

Thompson and Davy both observed that when energy changed forms, any matter indirectly involved in the process remained fixed in both type and amount. Shortly thereafter, James Prescott Joule (1818–1889) demonstrated that the energy itself, although altered in character, remained unchanged in quantity. Using elaborate devices consisting of paddles turning in barrels of water, he showed how X amount of heat produced Y amount of motion and how Y amount of motion could then once again be used to produce X amount of heat. Energy went back and forth undiminished among various identities in much the same way as matter did when it underwent chemical changes.

It didn't take long for the investigators of energy to realize that heat and motion weren't the only two forms it could take. Sound, light, electricity, and magnetism were all identified as being different aspects of the greater phenomenon of energy. And it was chemical energy that drove atoms and molecules to unite, disband, and rearrange themselves. All of these kinds of energy could, either directly or indirectly, be transformed into any of the others. Friction turned motion into heat, fire turned chemical energy into heat and light, engines turned heat into motion, and so forth and so on.

As all these interconversions were studied, described, measured, and related to one another it became increasingly obvious that energy, much like matter, could neither be absolutely created nor finally destroyed. Any ostensible source of new energy simply provided energy in a new format from other, already existing, supplies. Any apparent consumption of energy merely used up existing energy of one type as it generated an equal amount of energy of other types.

Lavoisier had demonstrated that matter, while undergoing astonishing and repeated changes, always remained matter and always remained unaltered in amount. This was called the law of the conservation of matter, and it held true for any isolated system, be it a stopped-up flask or the universe as a whole. (See Unit I.5.) The work of Joule and his successors resulted in a parallel law for energy that stated that within any defined and closed system the total amount of energy remained constant regardless of how it was transformed, rearranged, used, or stored. The creation of a closed energy system was far more difficult than stopping up a flask, but the law of the conservation of energy was repeatedly demonstrated to be valid. In the ultimate closed system as well—the universe—the sum total of energy, like the sum total of matter, was assumed to be finite and fixed.

Energy, once understood as something separate from matter, endlessly alterable in form but decidedly finite in amount, was subjected to intense investigation and exploitation. Motion was defined and studied as mechanical energy. The development and use of heat energy was formalized as thermodynamics. Electricity and magnetism were rapidly domesticated. While the raw ingredients of matter were synthesized into substances unknown in nature, the raw forces of energy were captured, channeled, and unleashed in novel ways. Indeed, the control of matter and the control of energy became like two sides of a single coin. Matter, to be changed, required the use of energy. Energy, to be useful, had to produce desirable changes in matter. The physical world, long held to be a purely material construction, was seen instead to be a vast, continual interplay between matter and energy.

The laws of the conservation of both energy and matter have since been subjected to revision due to the discovery of nuclear energy. By converting matter into energy nuclear reactions violate both laws. (See Unit II.29.) But within the realm of more conventional, nonnuclear events and processes these two laws remain valid and powerful tools and are largely responsible for the Industrial Revolution and the avalanche of inventions, gadgets, and gizmos that have followed it.

II.3 TRANSFORMATIONS OF ENERGY

In addition to its various forms, such as heat, motion, sound, and light, energy is capable of existing in two different states. The first of these, like the solid and liquid states of matter, is fairly obvious. The second, like matter's gaseous phase, is pretty hard to identify unless you know what you're looking for.

The readily discernible state of energy is its kinetic or active phase. The motion of a cannon-borer, the glow of a candle, the wail of an electric guitar, and the warmth of an electric blanket are all kinetic states of energy. Energy in its active phase is detectable by the human nervous system, usually as motion, heat, light, or noise.

The other state of energy is its potential or hidden phase. Energy in its potential phase lies dormant and unused, but possesses the ability to become active energy in the future. There's potential energy in a piece of wood because the wood is capable of being used as fuel sometime later on. The chemical energy hidden in the wood will be liberated as kinetic heat and light when it burns. There's potential energy in the wound-up mainspring of a watch because of what will happen over time. As the spring unwinds the potential energy will be slowly converted to the mechanical motion of the hands. The potential energy in a battery is capable of being released as electrical energy that can in turn be transformed into other forms of kinetic energy: light in a flashlight, noise in a radio, heat in a handwarmer, or motion in a child's toy.

Both states of energy are included in the law of the conservation of energy. Energy can and does move back and forth between its potential and kinetic phases without any loss of quantity. The classic example of this is a pendulum. While swinging from one end of its arc to the other a pendulum displays kinetic energy of motion. At the end of its swing, however, it pauses. At this instant all its kinetic energy has been expended to elevate it as far as possible. At the top of its swing a pendulum contains only hidden energy, known as potential energy of position. All this means is that it's capable of falling. After a momentary pause it immediately does so. Its potential energy of position reverts to kinetic energy of motion as it swings once more and the cycle repeats.

Energy, then, isn't only constantly changing from one form to another but from one state to another as well. As Thompson and Davy first realized, one form of kinetic energy can be transformed into another. Mechanical motion can become heat. As the pendulum demonstrates, one state of energy can be transformed into another. Potential energy becomes kinetic energy and vice versa. During each and every change of form or state the law of the conservation of energy holds true. The energy after the change equals the energy before the change.

Sometimes this is easy to see, sometimes not. A cannon-borer and a pendulum are pretty straightforward examples. But how about a glass of Kool-Aid perched on the edge of a kitchen counter? The glass, like a pendulum at the top of its arc, has potential energy of position. It can fall. When bumped by a curious and/or thirsty child it does so. Its potential energy of position becomes kinetic energy of motion—straight down. But what happens to the energy when the glass hits the linoleum and motion ceases? Well, some of it becomes noise. Rarely does a glass shatter without making noise of some kind, and noise is a form of kinetic energy. Some of it remains as motion. The Kool-Aid splashes and the glass fragments scatter across the kitchen.

Fine. But where does the energy go after the noise has died away and the splash and the scatter have concluded? Good question. The answer is heat. Noise, when it dissipates, is absorbed by surrounding walls, floors, or what have you and transformed into heat. The incidence of noise on a surface actually raises its temperature, although very marginally. (See Unit II.16.) The liquid and the glass shards also give up their kinetic energy of motion as heat. As they slide across the floor friction slows and finally stops them, and friction turns motion into heat, just as Thompson noted. (See Unit II.8.) The linoleum is warmed as surely as were the Bavarian cannons.

Almost any time energy appears to disappear, almost any time the law of the conservation of energy appears to be violated, the explanation is heat. The noise energy of a church choir, a clap of thunder, or a food processor isn't lost when silence returns, but is transformed into undetectable amounts of heat.

The light energy of a flashbulb, a stroke of lightning, or an "Open 24 Hours" sign isn't lost when darkness resumes, but is transformed into minute quantities of heat. The mechanical motion of a braking car, a falling tree, or an avalanche isn't lost when motion ceases, but is transformed eventually into heat.

Heat represents a sort of primal condition of energy. It's from a vast reservoir of heat that other forms of energy are derived, and it's into heat that all other forms of energy eventually deteriorate. Heat, in fact, is involved in every single transformation of energy. The heat of the sun is the starting point for nearly every use of energy you can think of. The sun directly provides the energy that makes plants grow, water evaporate, the wind blow, rain fall, and rivers flow. It's also the sun that fueled the growth of those prehistoric organisms that have since decayed into coal and petroleum over the years. This energy of the sun stored long ago now enable jets to fly, cars to roll, fires to burn, and factories to operate.

Even energy transformations that seemingly have nothing to do with heat involve it in one way or another. The pendulum is apparently a direct and total trade-off between kinetic energy of motion and potential energy of position. But a pendulum eventually stops swinging. The reason it stops is that part of its energy is continually being drained away as heat. Friction at its pivot point, as well as with molecules of the surrounding air, slowly and surely turns all its mechanical energy into heat.

No transformation of energy, in fact, takes place without the production of at least a little bit of heat. This heat is usually undesired and counter-productive. One hundred units of electricity won't yield 100 units of light because a few units will be lost as heat during the exchange. One hundred units of motion won't yield 100 units of noise for the same reason.

All machines, all appliances, all manufacturing processes, and virtually every other man-made or natural event requires the transformation of energy, and all of them are less than 100 percent efficient because of the unavoidable production of heat. It's the inherent inefficiency of energy transfers that prohibits the construction of a perpetual-motion machine. The motion of any machine, like that of a pendulum, will eventually become heat, and motion will eventually cease without the input of energy from an external source.

II.4 MECHANICS

Long before anything as sophisticated as a perpetual-motion machine was ever dreamed of, much less proven to be theoretically impossible, simple machines were invented and effectively employed. Long before the kinetic and potential states of energy were identified and shown to be interchangeable, the lever, the wheel, the bow, the inclined plane, and their derivatives were used to hoist loads, transport materials, hurl arrows, and erect buildings. Long before the law of the conservation of energy was discovered, the basic problems involved in relocating the rearranging matter were the subject of serious thought and tremendous effort.

Long before energy was even recognized as something nonmaterial, what's now known as mechanics—matter in motion—was systematically investigated for a number of reasons. For one thing, it was of absolutely vital importance. Survival depended to a great extent on one's ability to modify the immediate environment and to shield oneself from the greater external environment. That meant physically manipulating stuff into new sizes, shapes, and locations. Secondly, mechanics was the most plainly evident form of energy. Electricity and magnetism were generally unknown, sound and light totally beyond comprehension. Heat, in the form of fire, was obvious enough in its effects, but the manner of its operation was unfathomable. Mechanics, though, involved relatively familiar entities—human labor and solid stuff.

The Greeks, naturally, were the first to make the study of mechanics an academic exercise. While their predecessors and contemporaries pushed, pried, and pulled randomly looking for optimal results in hurling rocks or building pyramids, the Greeks tried to figure out exactly where to push or pull based on some form of reason. Their skills in the arts, architecture, and warfare were founded largely on theoretical knowledge of basic mechanical principles. Their expertise in this area is personified by Archimedes (c.278–212 B.C.), whose first love was mathematics but whose most valuable contributions to posterity lie in the area of engineering. Archimedes formulated many laws of mechanics that still endure and applied them to practical problems. He's best remembered for his justified claim that, given a big enough lever and a place to stand, he could move the world.

But it was Aristotle again rather than Archimedes who, initially at least, had a greater, longer-lasting impact on the study of mechanics. Matter, besides consisting of four elements and four basic properties, said Aristotle, was subjected to two distinct types of motion: natural and violent. Natural motion was vertical and the result of an object's content of either gravity or levity. Rocks had gravity and went down. Fire had levity and went up. Air contained equal parts of gravity and levity and didn't go anywhere.

Any nonvertical motion was considered a violent and temporary disturbance of the preordained operation of gravity and levity. Violent, horizontal motion required an unnatural push of some kind for as long as it continued, and ceased immediately when the interfering push ceased. An arrow in flight was a clear infraction of natural motion. As long as it traveled horizontally it was assumed to be under the influence of unnatural forces. In this case it was assumed to be the air that accelerated the arrow, carried it along its path, and finally slowed it to a standstill. It was the air, and not the bow, that provided the power, because only the air had continuous contact with the arrow and therefore only the air could exert any force upon it. Once the air tired and the unnatural force ceased, natural motion took over and returned the arrow vertically to its proper position on the earth's surface.

■■■■■■■■■■■■■■■■■■■■■■■■■■■■■■■■■■■■■■

Archimedes

Archimedes was successful both as an inventor and as a mathematician. Among his inventions the most notable was a device used to raise water for irrigation. It's used to this day in some parts of the world and known as Archimedes' screw. His mathematical accomplishments included the development of a number system, the power of which he demonstrated by calculating how many grains of sand it would take to fill the universe. He's also anecdotally remembered for his shout of "Eureka!" (I've found it!) upon discovering how to determine the purity of King Hiero's gold crown.

Archimedes was a native and lifelong resident of Syracuse, Sicily, then a Greek colony. There he was a close friend of King Hiero and helped defend the island against the Romans. The war machines he devised, most notably the catapult and grappling hook, prolonged the siege for several years. When the Romans finally overran Syracuse they found an aged Archimedes drawing geometric figures in the sand and slew him on the spot. His final words were reportedly, "Don't disturb my circles."

■■■■■■■■■■■■■■■■■■■■■■■■■■■■■■■■■■■■■

The theory was a pretty dismal one. It gave to the air the dual and opposing duties of transporting, then detaining, the arrow. It gave to the arrow a flight path like that of the water going over Niagara Falls. To the archer and the bow it gave no function whatsoever. It was so bad that, unlike many of Aristotle's other false starts, it failed to survive even the intellectual bankruptcy of the Dark Ages.

Medieval scholars, despite their deficiencies in creative thoughts and their steadfast loyalties to traditional knowledge, found the mechanics of Aristotle irrelevant to everyday experience. Too timid to trash the whole theory, they added to it something called impetus, a ghostly substance of the phlogiston/caloric genre that flowed into and out of objects thereby imparting to them and taking away from them motion as it did so. Aside from providing some new terminology, the creation of impetus did little to advance the understanding of matter in motion.

II.5 | INERTIA

Despite their rather obvious absurdities, archaic ideas about how an arrow traveled through the air or how a stone fell to earth remained unchallenged until that fateful day in Pisa when Galileo dropped a couple of unequal weights side by side. (See Unit I.3.) The notion of differing rates of descent for objects with differing amounts of gravity was an integral and central part of traditional thought regarding objects in motion. When Galileo demonstrated that when one ignores the effects of air resistance all objects actually fall at the same rate he exposed once and for all the fallacies of Aristotle's gravity and levity, natural and violent motion, and impetus. With one scientifically conducted experiment he destroyed forever an erroneous supposition that had been carefully fostered for centuries.

Constructing a proper replacement for what he had so recklessly disposed of turned out to be a bit more difficult. Galileo's first task was to find out just how objects really did behave when they fell. He knew that all objects fell at the same rate, but what was that rate? And did it change during the course of a free-fall or remain the same?

With a series of simple but ingenious experiments Galileo first proved that falling objects accelerated during the course of their descent. The longer they remained subject to the force of gravity the faster they traveled. This explained why a flowerpot falling from a third-story balcony struck whatever it struck with more force than one tumbling from a second-story balcony. It was the extra speed, gained during the extra time in the clutches of gravity, that gave the third-story flowerpot its extra force of impact.

This fairly rudimentary finding, obvious to anyone who has ever fallen from several different heights or contemplated the consequences of doing so, was a bit of a revelation. The Aristotelian/impetus school of thought had always held that a constant, steady force resulted in a constant, steady motion. Now here was Galileo proving that a constant, steady force—and what could be more unwaveringly applied than gravity?—resulted not in constant, steady motion, but in constantly *changing* motion. Specifically it resulted in constant, steady acceleration.

He next demonstrated that in the absence of any sort of interfering force motion will continue unchanged in either direction or speed. The totally original, utterly revolutionary notion that motion is self-sustaining seemed at the time an absurdity but is now the very cornerstone of mechanics. It's called inertia.

Inertia is the tendency of bodies at rest to remain at rest and the tendency of bodies in motion to remain in a state of steady motion. Inertia is what keeps the dishes on the table as the magician snaps the tablecloth out from under them. The dishes, not directly subjected to the yank of the magician, tend to remain where they are. Inertia is what what keeps your bowling ball rolling down the alley after it's left your hand. The bowling ball, even though no longer being propelled by your delivery, tends to maintain its course and speed. In the absence of any counterforces linear motion proceeds eternally once commenced.

Galileo's inertia was the first breakthrough in the study of mechanics and the first thing that logically explained certain aspects of matter in motion. An arrow wasn't accelerated by any magical property of the air, but by the

● ●

Falling Bodies

Galileo was, it seems, interested in everything. As a child he engineered elaborate mechanical toys and excelled in both art and music. He later studied medicine but, for financial reasons, ended up teaching mathematics at the University of Pisa. It was there that he reasoned that two bricks cemented together should fall no faster than the same two bricks falling separately side by side. His subsequent disproof of Aristotle caused such an uproar among the faculty that he was forced to resign.

It was a few years later, at the University of Padua, that he was able to study the behavior of falling bodies in more detail. The task was complicated by the fact that there existed at the time no instruments capable of accurately measuring the particulars of an object's speed, position, and elapsed time of descent while in a free-fall. Galileo cleverly circumvented this obstacle by studying balls as they rolled down ramps. He thereby "diluted" gravity to the point where its effects could be faithfully observed.

Galileo quickly determined that as a ball rolled down a ramp it gained speed. Further experimentation revealed that the acceleration was constant and predictable, a logical consequence of the persistent, unwavering force of gravity. This behavior, he correctly concluded, applies to freely falling bodies as well as to downward-rolling ones.

—Does a penny dropped from a fourth-story window take twice as long to hit the ground as one dropped from the second floor? Why or why not?

—Could a penny, dropped from a tall-enough building, be as dangerous as a bullet fired from a gun?

Galileo proceeded to roll balls down one ramp, across a level surface, and then back up another ramp. During the horizontal portion of their journeys, he found, the balls neither sped up nor, discounting the effects of friction, slowed down. On their way back up the ramps they gradually slowed and then stopped as they fought the downward tug of gravity.

—How did the final height achieved by the balls on the up ramp compare to the height from which they had started out on the down ramp? Would changing the angle of the ramps make any difference?

—How does the speed at which a stone is thrown upward compare to the speed at which it strikes the ground?

● ●

bowstring during the brief moment it was applied. Once in flight an arrow was carried along by its own inertia, not impetus from the air. An arrow in flight didn't speed up, then slow down, but, except for the slowing effects of wind resistance, traveled at constant horizontal velocity. While the arrow traveled horizontally it also accelerated downward under the steadily applied force of gravity. The flight path of an arrow traced out an interaction between constant horizontal motion and accelerating downward motion. It was a complex curve, a parabola to be exact, and not a waterfall.

With the workings of gravity and inertia better understood it became apparent that there was no major difference between natural and violent motion. All motion, whether horizontal or vertical, evidently remained constant when left alone and changed when subjected to a force of some sort, be it the tug of gravity or the snap of a bowstring. It was the interaction of various forces acting upon it, not its content of gravity and/or levity, that determined an object's motion, position, and acceleration. It was inertia, not impetus, that laid at the heart of mechanics.

By the time Galileo died, in 1642, he had not only effectively swept away the erroneous theories of the past, but had identified and collected most of the components necessary to construct a truly functional system of mechanics. All that remained was for a master architect to assemble them.

II.6 THREE LAWS OF MOTION

Later that same year, 1642, there was born in England a son to Mr. and Mrs. Newton. They named the boy Isaac (1642–1727). He was to grow up to become one of the greatest thinkers the world has ever known. The coincidence of Galileo's death and Newton's birth is of no real consequence, but nicely illustrates the fact that one's work began where the other's left off. Galileo razed the ramshackle mechanics of Aristotle and the Middle Ages, cleared the way for a more sensible structure, and assembled the materials with which to build it. Taking advantage of Galileo's preparations, Newton erected virtually the entire framework of modern mechanics.

Relying more on sheer intellect than inclined planes, using calculation rather than experimentation, Newton organized the work of Galileo, along with a mountain of assorted other observations and data, into a cogent, cohesive, and comprehensive set of three laws and the formulas derived from them. This masterpiece of mechanics, which endures pretty much intact to this day, he published in 1687 as the *Philosophiae Naturalis Principia Mathematica*. Newton's three laws of motion are as follows:

1) A body remains at rest or, if already in motion, remains in uniform motion with constant velocity in a constant direction, unless it's acted upon by an unbalanced external force.

2) An unbalanced external force acting upon a body at rest or in uniform motion accelerates that body in the direction of the force and to a degree directly proportional to the magnitude of the force.

3) Whenever one body exerts a force on a second body the second body exerts a force on the first body. These reciprocal forces are equal in magnitude but opposite in direction.

Although no one before Newton had discovered the three laws of motion, everyone had dealt with them on a daily basis. You have too, whether you realize it or not. Newton, by publishing the *Principia,* did nothing to alter the way the world worked. He merely explained it in precise, mathematical language.

Newton's first law is a restatement of Galileo's inertia. Rest and uniform motion represent states of equilibrium and remain as they are unless or until disturbed by an interfering force of some kind. What could be more plainly true? Your car, parked outside the post office, stays parked there unless or until disturbed. Your car, cruising down the highway, remains cruising down the highway until or unless a hill or friction slows it, a jab to the accelerator speeds it, a bridge abutment stops it, or a twist of the steering wheel or an irregularity in the road diverts it. Everything remains as it is unless or until something takes place to cause change. That, in a nutshell, is Newton's first law.

Newton's second law deals with change or the disruption of inertia by an external force of some sort. Any such disruption, according to the second law of motion, takes place in the direction and to the extent that an external force is applied. This, too, is just good old common sense. Anything, be it a tank or a Tinker Toy, responds to a shove by moving in the direction of that shove and to the degree shoved. The force of its engine changes the speed of a car, and hence its inertia, by accelerating it from zero to 10 kph, from 10 to 20 kph, or from 100 to 110 kph. The greater the force applied by the engine the greater the change in speed. The force of its brakes slows a car from 110 to 100 kph or from 10 to zero kph. The extent of slowing, again, depends directly upon the amount of force applied. And a car's course is altered in the direction and to the extent of a force applied via its steering mechanism or a crosswind. In each case the direction of the force determines the direction of the change and the size of the force determines the magnitude of the change. That is Newton's second law.

Newton's third law says every force is exactly offset by a counterforce acting in the opposite direction with an equal magnitude. Or, more popularly, to every action there is an equal and opposite reaction. This law is less obvious than the first two, but equally valid. When a hammer strikes a nail the nail applies as much force to the hammer as the hammer does to the nail. When a bullet strikes a wall two equal forces, going in opposite directions, are applied—one by the bullet on the wall, the other by the wall on the bullet. And when the gun was fired a similar pair of forces was generated. One side of the force equation propelled the bullet out the end of the rifle, while the other side forced the butt of the rifle into the shoulder of the gunman. No force takes place in isolation; each is mirrored by an equal and opposing force. That is Newton's third and last law of mechanics.

II.7 THE CONSERVATION OF MOMENTUM

The central concept of so-called Newtonian, or classical, mechanics is that of momentum. Momentum, at its simplest, is a quantitative expression of inertia. It's the measure of the strength of a body's tendency to remain at rest when at rest or to remain in uniform motion when in motion. It's the measure of how much a body will react to any particular external force. And it's the measure of how much force one body in motion will exert on a second body and, in turn, how much force the second body will exert on the first.

The momentum of any object is the direct function of two, and only two, things: its velocity and its mass. An increase in either of these two properties will result in a corresponding increase in its momentum. A fired bullet possesses more momentum than a thrown one due to its greater velocity. It has a greater tendency to maintain its forward motion, is harder to deflect, and causes greater damage to whatever it strikes. A fired artillery shell, on the other hand, has more momentum than a fired bullet due to its greater mass. It's more resistant to external forces and has a more powerful impact on whatever it strikes.

Both the bullet and the artillery shell not only exert force on what they hit, but receive force as well. In fact, whatever force they exert, as determined by their content of momentum, will be exactly matched by the force they receive. These reciprocal forces, being equal in amount but opposite in direction, cancel one another as far as momentum is concerned. According to Newton's third law of motion, in fact, the total amount of momentum existing after a gun or a cannon is fired is exactly equal to what it was before the act took place. Within any arena of action, including the universe as a whole, there's a finite and constant amount of momentum. Momentum, much like matter and energy, can be redistributed but can never be absolutely created or destroyed. This, as you may have already guessed, is the law of the conservation of momentum.

The perfect illustration of this law is a rocket floating silently in space, free of the clutches of gravity and connected to nothing—its engine shut down. In this state it has a certain amount of momentum based on its mass and velocity. It's in uniform motion and has a tendency to remain as it is—to retain its current course and speed—according to Newton's first law.

All this changes when the engine's ignited. The combustion of fuel within the rocket produces a sustained explosion of exhaust. Were this explosion to take place out in the open the resultant exhaust particles would be radiated

● ●

Thrust

Imagine you are a rocket engineer. Your goal is to build rockets that go ever farther and faster. You are, in other words, always looking for ways to give your rockets more forward momentum, what in your business is usually referred to as thrust.

Your constant adversary in this endeavor is the law of the conservation of momentum. This unbreakable law of nature states that the forward momentum, or thrust, of any rocket is determined by one and only one thing—the rearward momentum of its exhaust. The rearward momentum of its exhaust, like the momentum of anything else, depends solely upon its mass and its velocity.

There are, then, only two ways for you to boost the thrust of your rockets: you can increase the amount of exhaust spewing out their tails, or you can increase the speed at which their exhaust is spewed.

Increasing the mass of the exhaust, you realize, is a somewhat self-defeating strategy. The mass of a rocket's exhaust is strictly limited to the mass of fuel it burns and, because there are no filling stations in space, a rocket must carry all its own fuel. Carrying extra fuel means building a larger, heavier rocket that in turn requires more thrust and more fuel.

What you really want to do, then, is increase the velocity of your rocket's exhaust. By doing so you can squeeze additional thrust from any given amount of fuel. Your attempts to increase exhaust velocity generally consist of formulating fuels that burn very hot and fast. You also design special nozzles that concentrate and accelerate exhaust gases. You're also experimenting with nuclear engines that, because of the intense heat they generate, promise to be two to three times more efficient than conventional combustion when perfected.

—How do you slow down and steer your rockets?

—What are the advantages of building rockets in stages? Why are the first stages so much bigger than the later ones?

—Is there any advantage to launching a rocket from a mountaintop?

—In fact, it takes less thrust to launch a rocket from a launching pad near the equator that it does from one near either of the poles. Why?

—Why do astronaut-carrying rockets lift off so slowly compared to those used to shoot down airplanes or deliver bombs?

● ●

symmetrically in all directions—north, south, east, west, up, and down. Their individual momenta would cancel out and, mathematically speaking, reduce to what they were before the explosion. The motion of the rocket, being subjected to equal amounts of force in every direction, wouldn't be affected.

But the steady explosion of exhaust doesn't take place out in the open. It takes place inside a chamber that is designed to concentrate and expel all the exhaust in one direction—out the tail. The force exerted on the rocket isn't symmetrically balanced then, but unbalanced. The rocket, subjected to an unbalanced force, undergoes a change in its motion according to Newton's second law. As the exhaust heads south the rocket heads north. The momentum involved has been redistributed, but its total hasn't been altered in the least.

A jet airplane propels itself the same way. The only difference is that the jet relies on the surrounding atmosphere for the oxygen needed to burn its fuel while a rocket carries along with it everything it needs. The exact same thing makes a released balloon snake through the air, a loose firehose whip around, or a lawn sprinkler twirl. In each case the reaction is the result of an action of equal magnitude but opposite direction. The motion of the balloon depends on the momentum of the expelled air, and the firehose and the sprinkler depend on the momentum of the expelled water. Neither the rocket, the jet, the balloon, the firehose, nor the sprinkler require anything external in order to operate. They operate solely by redistributing the momentum they already possess. You can experience the same effect yourself by throwing an anchor over the side of a rowboat. Whatever oomph you supply to send the anchor one way will be matched by a counteroomph that rocks the boat in the opposite direction.

The redistribution of momentum in accordance with the law of the conservation of momentum isn't always so apparent. When a bullet is fired from a gun the forward burst of the bullet is noticeably matched by the backward kick of the gun. Should the bullet strike a tin can it's also obvious that the momentum lost by the slug is gained by the can. But what happens when the bullet strikes the side of a mountain? The bullet is stopped in its tracks but the mountain just sits there. What happens to the momentum lost by the bullet?

The law of the conservation of momentum tells us that it can't just vanish, but must be redistributed. Where does it go? It goes the only place it can go—into the mountain. The mountain, along with the planet to which it's attached, has an enormous ability to absorb and supply vast amounts of momentum without itself undergoing any perceptible change. The earth, in fact, serves as huge reservoir of momentum ready to provide it or store it as needed, which is quite frequently, as nearly all familiar events are eventually grounded in one way or another.

II.8 FRICTION

Objects at rest tend to remain at rest. Objects in motion tend to remain in motion. This is the concept of inertia first realized by Galileo and then formalized by Newton as the first law of motion. It takes more than inertia, though, to keep a car traveling down a highway or a jet plane flying across the sky. Both these motions, if they're to remain uniform, require a constant application of force. Without the constant input of force supplied by their engines a car slows and finally stops, a jet slows and falls from the sky.

Do cars and jets violate the first law of motion? Of course not. Motion remains uniform only in the complete absence of any unbalanced external force—only under ideal conditions. Cars and planes, though, along with the rest of the world's moving objects, operate under real conditions. They're all subject to an unbalanced external force that tends to slow and eventually stop them. They're all subject to friction.

Friction results whenever anything moves while in contact with anything else. Any object that moves makes contact with its immediate environment and experiences the slowing force of friction as it does so. A car is slowed by friction between its tires and the road, by friction among its internal moving parts, and by friction between its bodywork and the air. A plane, a Frisbee, and a golf ball are similarly slowed by the air. A boat is slowed by the air, too, as well as by the water in contact with its hull. A submerged submarine is slowed by the sea surrounding it. In order to maintain a constant speed each of these objects must receive a perpetual input of motivating force equal in magnitude to the slowing force of friction but opposite in direction. Without such a motivating force all will slow until they stop.

The force of friction opposes virtually every action. Friction increases the difficulty of opening a window or a jar of mustard, the turning of a doorknob or a steering wheel, grating cheese, sweeping a floor, mowing a lawn, shoveling snow, chopping wood, or brushing teeth. When acting contrary to deliberate or desirable actions in this manner friction is a tenacious and tireless adversary—an adversary that pilfers energy and lowers efficiency.

But friction, like almost anything else, has a few redeeming qualities as well. As it resists the motion of one object against another it acts as a sort of universal glue. As you walk, friction holds your right foot in place while you shove off of it and swing your left foot forward. Friction also holds your shoe onto your foot and your shoelaces in your shoe. Friction allows you to grab and control a jar of mustard, a doorknob, a steering wheel, a broom, or a toothbrush. Friction prevents an opened window from immediately closing again and allows you to stack up chopped firewood or pile up shoveled snow. Without friction nearly all our daily activities would be rendered comical if not impossible. A car's wheels would spin futilely without friction. Nails would slide out of beams, nuts would spin freely off of bolts, woven fabrics would unravel, and pedestrians would skid uncontrollably. The world would be an unimaginably slippery place without friction.

The energy that friction embezzles from mechanical motion ends up as heat. This is the fact that made such a historical impression on Thompson as he was boring cannons. (See Unit II.1.) Generally speaking, the heat produced by friction is a highly undesirable commodity. Heat born of friction wears out tires, shoes, and engines. Heat born of friction overheats machines, factories, and the workers therein. An ill-fitting shoe causes friction as it rubs back and forth, and the resulting heat gives rise to blisters. But friction-produced heat, like friction itself, isn't all bad. Thousands of fires, kindled by everyone from cavemen to Cub Scouts, have been started by the heat generated when two sticks are rubbed together. In more modern times matches and cigarette lighters rely on the same principle. And a pair of cold hands, like Davy's ice cubes, can be warmed by mutual friction.

II.9 | THE NATURE OF HEAT

As history suggests, motion and heat are closely related. It was the conversion of motion to heat that first alerted Thompson to the fact that energy existed apart from matter, and it was the quantification of that conversion by Joule that eventually led to the law of the conservation of energy. We now know that the basis of this relationship is the fact that heat is nothing more than motion at the molecular level.

The kinetic molecular theory of matter holds that all matter is composed of tiny particles called molecules, and that each and every one of them is in constant motion. As more and more heat is applied to matter this molecular mayhem accelerates. As matter cools its molecules slow down. Heat is nothing but molecular motion and vice versa. Friction is just a means whereby motion on a macroscopic scale is transformed into motion on a microscopic scale.

Heat, according to this description, can't exist apart from matter. Without molecules of matter to vibrate, heat is meaningless and impossible. This doesn't mean, as believed by everyone from Empedocles to Lavoisier, that heat is a material thing. No, heat is a form of energy, not matter. But heat can only be manifested through material means. A vacuum, an expanse of space void of matter, can't contain heat because it has no molecules available to vibrate.

Once the nature of heat as molecular motion is properly grasped it becomes immediately evident that cold is nothing but the absence of heat. Cold isn't a physical phenomenon in and of itself. Just as darkness is the lack of light, slowness the lack of speed, and silence the lack of noise, cold is the lack of heat. Properly speaking, a frozen burrito doesn't have more coldness than a warm one, it has less heat.

Heat, as already pointed out, can be produced from and turned into just about any other form of energy according to the law of the conservation of energy. Mechanical energy becomes heat readily via friction, and any sort of fuel-burning engine turns heat back into the mechanical energy that powers planes, trains, automobiles, lawn mowers, chain saws, and pumps. Heat, via the mechanical intermediary of a generator, is also a common source of electrical energy. Electricity, in turn, is commonly transformed into heat by all sorts of electrical appliances. Hair dryers, stoves, toasters, clothes dryers, percolators, deep fryers, soldering guns, water heaters, and corn poppers all turn electricity into heat, usually by forcing it through thin, highly resistant wires. As electricity flows through these wires it causes a sort of friction at the molecular level that produces the heat.

Chemical energy is another energy form that readily exchanges identity with heat. Almost any chemical reaction, in fact, can be classified as one of two types, depending upon its relationship with heat. Endothermic reactions are those in which heat is absorbed and stored as chemical energy. Exothermic reactions are those in which chemical energy is liberated as heat.

Endothermic reactions, generally speaking, are those in which simple substances are combined with the aid of heat to form more complex ones. The synthesis of nearly any artificial substance, like glass or rubber, requires a great deal of heat. Chemistry, in these reactions, goes uphill. The process involved prefers not to take place and must be pushed along with heat and often pressure as well. Plants use natural endothermic chemistry to transform simple nutrients from the soil, the air, and water to produce complex carbohydrates and proteins. This process, called photosynthesis, obtains its needed heat from sunlight.

Exothermic reactions, in contrast, can be thought of as running downhill. These reactions often take place spontaneously or continue of their own accord once given a shove in the right direction. Exothermic reactions give off heat as a complex substance is broken down, decays, or degenerates into simpler, more stable substances. The most common type of exothermic reaction is called oxidation, a process whereby old substances are rearranged into new ones by the action of oxygen.

Decay is a typical oxidation reaction in which the complex substances built up by photosynthesis deteriorate spontaneously into simpler substances. Although not always noticeable, the heat energy consumed during the building up is liberated during the breaking down, sometimes enough of it to trigger spontaneous combustion. Rust is another common oxidation process. As metals rust molecules of pure metal combine with oxygen in the air to form metallic oxides. Rust, like decay, is an exothermic reaction that proceeds with little or no encouragement. Indeed, great measures are taken to prevent rust from happening, often in vain. Rust, too, liberates heat. The heat isn't detectable because the reaction takes place so slowly. Nevertheless, it's there.

When an oxidation reaction proceeds rapidly enough to produce easily detectable amounts of heat and light it's called combustion. A tree combines water, carbon dioxide, and energy to form wood via photosynthesis. Combustion reverses this process to get water vapor, carbon dioxide, and energy from a piece of wood. Some metals also combust. A few of them do so violently. Lithium must be kept submerged in an oil bath because any contact with the air will cause it to burst into bright, hot flames. Other metals need high ignition temperatures, but will also eventually burn.

All metals oxidize in one way or another, either by rusting or combusting. Wood eventually oxidizes either by burning or rotting. As far as the transfor-

■■■■■■■■■■■■■■■■■■■■■■■■■■■■■■■■■■■■■

Rust

The degree to which metals rust varies greatly from one to the next. Some, most notably iron and steel, rust thoroughly. Others, like silver and aluminum, rust only superficially. The initial surface rust in these latter cases actually acts as a protective film of metallic oxide that insulates the rest of the metal from the air and thereby prevents further oxidation. Polishing these metals removes the dull layer of oxidation, reveals the shiny metal underneath, and exposes it to the air where, unless protected, it will then be oxidized.

The rusting of iron and steel generally requires the presence of water. The water doesn't directly participate in the formation of rust, but acts as a catalyst that allows oxygen and iron to unite chemically as ferric oxide, Fe_2O_3. The rusting process can be halted by denying the metal access to either oxygen or water. In theory, iron will not rust in water-free air or in air-free water. In practice, though, air usually contains water as humidity, water usually contains traces of dissolved oxygen, and iron eventually rusts.

■■■■■■■■■■■■■■■■■■■■■■■■■■■■■■■■■■■■■

mation of chemical energy into heat energy is concerned, it really doesn't make any difference. The same amount of heat will be produced either way. A log decomposing over a period of 20 years produces the same amount of heat as it would if it were burned over a period of 20 minutes.

II.10 | TEMPERATURE, NOT HEAT

A metal's preference to either rust or combust depends upon the temperature at which it's oxidized. A piece of wood likewise either rots or burns depending upon the temperature at which oxidation takes place. And it's temperature that determines whether a substance exists as a solid, a liquid, or a gas. Temperature, not heat.

There's a difference.

Heat is a quantitative measurement. It denotes an absolute amount of molecular motion. Molecule X, vibrating twice as enthusiastically as molecule Y, has twice the heat content as well. Similarly, molecules A and B, vibrating in identical fashion, together have twice the heat content that either has individually. Heat is cumulative. The total amount of molecular motion within any defined amount of matter is the amount of heat in that matter.

Temperature, in contrast, is a qualitative measure. It denotes the concentration or intensity of heat at any particular spot rather than the amount of heat in total. Very small amounts of heat, highly concentrated, can result in extremely high temperatures. Greats amounts of heat, vastly diffused, can exist at very low temperatures.

Heat and temperature often operate in unison. With a few important exceptions to be explained shortly, raising the amount of heat in a bathtub full of water also raises its temperature, and lowering its heat content also lowers its temperature. The concepts of heat and temperature in such a situation are closely linked, and the failure to distinguish between them is of no great consequence.

In other instances, though, the difference between temperature and heat is important. Consider a tubful of comfortable bathwater and a steaming-hot cup of coffee. The bath has a lower temperature but, at the same time, contains more heat. Even though the average speed of the molecules in the tub is slower than of those in the cup, the bath contains far more molecules than the coffee. The bathwater contains more heat than the coffee and has a greater capacity to warm the bathroom or the bather.

As the temperature of a sample of matter is raised the amount of heat per molecule increases. Every molecule vibrates more rapidly and more violently. Each needs a bit more space in which to shake about. A group of highly agitated molecules will spread out slightly in order to give themselves this extra elbow room. The group, as a whole, expands just a bit. As most objects get warmer, then, they also get bigger. The gaps between slabs of concrete in a highway are there to allow the concrete to expand under a hot sun. The force of expansion is so great that without the gaps neighboring slabs would crowd into one another, crack, and crumble. The expansion within most types of glass results in internal strains that cause them to shatter when subjected to rapid changes in temperature. Special kinds of glass, like Pyrex, are engineered to minimize and withstand such strains.

The expansion of heated substances, as well as being a nuisance, can also be taken advantage of. The cap on a ketchup bottle that simply won't budge can often be loosened by putting it under hot water for a few moments. The metal cap, when heated, expands more than does the glass of the bottle, grows a bit bigger relative to the bottle, and loosens. It's also the expansion of metals that operates most thermostats. When the metal within a thermostat cools below a certain point it shrinks enough to trip a switch that lights the furnace. When the metal is warmed up again by the running furnace its growth trips another switch that shuts down the furnace.

The expansion and contraction of substances due to temperature changes is nowhere more evident than in a thermometer. A thermometer, in fact, is nothing more than a gauge for measuring the volume changes of matter—usually mercury or colored alcohol—as it warms and cools. (See Figure II.1.) It was that old Italian troublemaker Galileo who first noticed the expansion that accompanied an increase in temperature, and it was Galileo who in 1593 capitalized on the effect to construct the world's first known thermometer, what he called a thermoscope. Based on observations and experiments with his

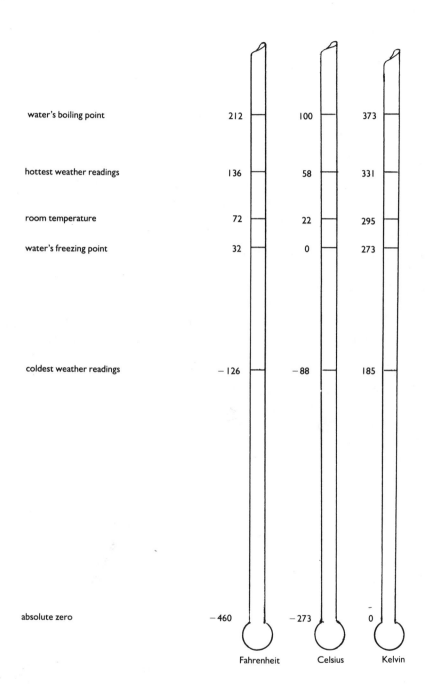

water's boiling point — 212 — 100 — 373

hottest weather readings — 136 — 58 — 331

room temperature — 72 — 22 — 295

water's freezing point — 32 — 0 — 273

coldest weather readings — −126 — −88 — 185

absolute zero — −460 — −273 — 0

Fahrenheit — Celsius — Kelvin

Figure II.1   Three temperature scales

new toy he successfully concluded that heat and cold weren't separate and opposite properties of matter, as Aristotle had held, but different aspects of a single phenomenon.

Galileo's first thermometer was understandably crude. It could tell you which of two objects was the hotter, but it couldn't tell you exactly, or even approximately, how *much* hotter. Worse, there was no way to compare the relative temperatures of two objects measured with different thermometers. This decidedly unscientific situation remained unchanged for the next hundred years. Thermometers of greater precision and reliability were developed, but the problem of standardization remained unsolved.

Then, in 1714, Daniel Gabriel Fahrenheit (1686–1736) developed a thermometer with universal reference points and a numerical scale, thereby bringing law and order to the world of temperatures. Any temperature could now be objectively measured and meaningfully expressed. A temperature of 1, 11, or 101 degrees Fahrenheit meant one and the same thing to everyone everywhere.

Still, there were those who remained dissatisfied. The most common complaint about the Fahrenheit scale was its casual treatment of the freezing and boiling points of water, probably the two most important temperatures in the world. To rectify these shortcomings a Swedish astronomer, Anders Celsius (1701–1744), proposed an alternative scale with the freezing point of water set at zero degrees and its boiling point at an even 100 degrees in 1743. Properly called Celsius but popularly referred to as centigrade, this temperature scale was quickly adopted as part of the metric system and has since become widely accepted along with it.

The scientific community, however, was still unhappy about the need to use negative numbers for all temperatures below the freezing point of water, which it regarded as a confusing and unnecessary inconvenience. Temperature was a continuum, a measure of the intensity of heat. There was no such thing as a negative concentration of heat, so there should be no such thing as negative temperature.

If all temperatures are to be positive there can be no temperature lower than zero. This means that zero on an all-positive temperature scale must represent the lowest temperature possible. There is such a temperature. Called absolute zero, it corresponds to a total absence of heat, a total lack of molecular motion. At absolute zero molecules stand motionlessly in place. Nothing happens. This coldest possible temperature works out to be about $-273°C$, or $-460°F$. This, then, is what's labeled as zero on the absolute or Kelvin scale. But the Kelvin scale, despite some obvious advantages over both the Fahrenheit and Celsius scales, is rarely encountered outside of academic circles. We continue to struggle with conversions between the two popular scales and deal with awkward and nonsensical negative degrees. The Kelvin scale is important, though, for it reminds us that there's no such thing as a negative concentration of heat.

Fahrenheit

Daniel Gabriel Fahrenheit was born in Germany, but later moved to Holland where he earned his living by designing and building weather instruments. He was familiar with Galileo's primitive thermoscope and a number of other temperature-measuring devices that followed it, but wasn't satisfied with any of them. He decided to see if he could come up with something better.

His first step was to shape a long, thin, hollow glass tube. One end of the tube he formed into a small bulb. This he filled with mercury, a silvery liquid metal. He then sucked the air out of the tube and sealed the other end.

Fahrenheit then plunged his invention into a freezing mixture of ice and salt several times. He noted that the level of mercury in the tube always settled at the same spot, which he marked with a scratch. He then took the body temperature of a healthy man, namely himself, and noted with a second scratch the higher level that the mercury reached.

He then proceeded to calibrate his instrument by measuring the distance between his two scratches and dividing it into 96 equal increments, what we now call degrees. The lower mark he labeled as zero the upper one, of course, 96. He continued to mark off uniform degrees beyond zero in one direction and beyond 96 in the other.

He then had a thermometer that anyone, anywhere, could re-create. Anyone who did so would find that pure water boiled at 212 degrees and froze at 32 degrees. Any temperature measured on a thermometer constructed by Fahrenheit's method could be meaningfully communicated to and faithfully duplicated by anyone else with a similarly constructed device.

Although other scales have been popularized since, Fahrenheit was the first to standardize the measurement of temperature.

—What special properties of mercury made it Fahrenheit's choice to fill his thermometer? Could he have used water?

—Fahrenheit originally divided his scale into only 16 rather than 96 intervals. What are the advantages of having more numerous, smaller degrees?

—What is the purpose of scattering salt on an icy sidewalk? Does this do any good if the temperature falls below 0°F?

—Body temperature, we now know, is 98.6°F, not 96°F. What could have caused Fahrenheit's error?

■■■■■■■■■■■■■■■■■■■■■■■■■■■■■■■■■■■■

Temperature Conversions

Fahrenheit had a bit of trouble with precision, incorrectly measuring his body temperature as 96 degrees rather than its true 98.6 degrees. Celsius, meanwhile, had some conceptual difficulties, initially and rather unconventionally proposing a boiling point of zero degrees and a freezing point of 100 degrees for water. Now that the two scales are accurate and going in the same direction, one can be converted to the other by the following formulas:

$$C° = 5/9 \ (F° − 32), \text{ or}$$
$$F° = 9/5 \ C° + 32$$

The Kelvin scale, named after Lord Kelvin, the title of British physicist William Thomson (1824–1907), proceeds from absolute zero by Celsius degrees. A Celsius temperature can therefore be converted to a Kelvin temperature simply by adding the number 273. Water freezes at 273°K and boils at 373°K.

All three of the scales, as well as any others that might be devised, are open-ended. Absolute zero represents the coldest temperature possible, but there's no theoretical limit for high temperatures.

■■■■■■■■■■■■■■■■■■■■■■■■■■■■■■■■■■■■

II.11 THE LATENT HEAT OF FUSION

Temperature is a measure of the intensity of heat at a certain location. It's an indication of the concentration of heat within a particular sample of matter. When dealing with only one object or collection of matter in isolation heat and temperature have a pretty straightforward relationship. When heat is added to a tubful of water, heat concentration as well as heat content increases. Temperature rises. Temperature moves in tandem with the heat content of the bathwater, rising as heat content rises and falling as heat content falls. Usually.

There are two important exceptions to this correlation between temperature and heat. These exceptions occur at those critical temperatures where solids turn to liquids and where liquids turn to gases; for example, at the freezing and boiling points of water. At its freezing point, be it labeled 32°F, 0°C, or anything else, the removal of heat from water no longer results in a lowering of its temperature. At its boiling point, on the other hand, water no longer gets any hotter as more heat is added to it. Instead of getting colder, the water freezes. Instead of getting hotter, the water boils. This is significant. At its freezing point water doesn't solidify *and* cool with the removal of heat, but solidifies *instead* of cooling. Likewise, at its boiling point, water doesn't vaporize *and* get hotter with the addition of heat, but vaporizes *instead* of getting hotter. The energy that would otherwise go into changing its temperature goes instead into changing its state.

Water at its freezing point will turn to ice as heat is removed from it. Its temperature, though, doesn't change during this transformation from the liquid to the solid phase. The same amount of water at the same temperature will therefore contain different amounts of heat depending upon whether it exists as a liquid or as a solid. The difference in heat content of liquid water and solid ice at 0°C is called the latent heat of fusion. It's called latent heat because it's hidden—it causes no change of temperature. To melt a piece of ice or freeze a quantity of water at 0°C it's necessary to add or remove, respectively, the latent heat of fusion. Water at its freezing point won't freeze until its latent heat of fusion is extracted. Ice at its melting point won't melt until its latent heat of fusion is supplied.

The involvement of the latent heat of fusion as water changes from a solid to a liquid or vice versa is more than a technicality. The latent heat of fusion is a rather sizable source of heat as water freezes, and an equally sizable consumer of heat as ice melts. In numerical terms, the transformation from the liquid to the solid phase of a given amount of water involves 80 times as much heat as does raising or lowering its temperature by 1 degree Celsius. To freeze a tray full of freezing-cold water to make ice cubes requires as much thermal effort as cooling the water in that tray by 80 degrees Celsius or the equivalent, 144 degrees Fahrenheit. This is why it seems to take forever to make ice cubes, even when the water is already icy cold. There's a payoff, though. The same heat of fusion that makes ice cubes so difficult to make also renders them difficult to melt. An ice cube effectively cools a glass of lemonade because as it melts it absorbs the required heat of fusion from its surroundings. The lemonade is cooled as it gives up its heat to the ice.

Citrus growers take advantage of the heat of fusion on a grander scale. When they anticipate a possible crop-destroying freeze they spray their orange trees with large quantities of water. As the cold air rolls in the water on the trees freezes, releasing its latent heat of fusion as it does so. The heat thus released stabilizes the air temperature at the freezing point. The juice in the oranges, being a solution, has a lower freezing point than that of pure water. The freezing water prevents the temperature of the air from falling to where it would freeze and damage the oranges.

The latent heat of fusion thus acts as a buffer between subfreezing and suprafreezing temperatures. Large quantities of heat can flow into and out of water at its freezing point without any change in temperature. To budge ice, water, or an ice-water combination from 0°C requires the addition or removal of the heat of fusion and then some. The freezing point is a stable temperature, resistant to change. A glass of iced lemonade remains at 0°C until every last crystal of ice is melted. Only then does it start to warm up. A tray of water remains at 0°C until it's frozen solid. Only then does the temperature of the ice start to drop.

II.12 THE LATENT HEAT OF VAPORIZATION

At the transition point between the liquid and solid states there exists an energy threshold called the latent heat of fusion. A similar threshold, called the latent heat of vaporization, exists at the juncture between the liquid and gaseous states. At its boiling point a great deal of heat must be supplied in order to turn a liquid into a gas. Conversely, the condensation of a gas into a liquid at the boiling/condensation point releases an equal amount of heat.

Typically, the latent heat of vaporization for any substance greatly exceeds its latent heat of fusion. Whereas 80 times as much heat is involved in melting or freezing a quantity of water as in raising or lowering its temperature 1°C, the condensation or vaporization processes require 540 times the energy involved in a 1°C temperature change. This is why it seems to take forever to boil a pot of water, whether watched or not. The heat needed to turn the liquid water into steam, once it has already reached its boiling point, is over five times as great as that needed to raise its temperature from its freezing point to its boiling point. Once the boiling point has been achieved the addition of more heat results in no additional rise in temperature, but in vaporization. (See Figure II.2.) A pot of water at a simmer on a back burner has the same temperature as one on the front burner boiling vigorously over a full flame. The extra heat going into the front pot merely speeds the rate of vaporization.

The latent heat of vaporization, like the latent heat of fusion, can be put to good use. A refrigerator provides the most common, practical, and graphic example of the latent heat of vaporization at work. The substance alternately vaporized and condensed in this case is a refrigerant, most commonly Freon, but the underlying principle is the same as if it were water flip-flopping between its gaseous and liquid states.

The Freon in a refrigerator constantly circulates within a long, convoluted loop of sealed tubing. Part of the loop lies within the food-storage compartment and part of it lies outside—usually behind—the appliance. The Freon enters the refrigerator proper under high pressure and as a liquid. Once inside, the pressure is reduced. The Freon, no longer forced into its dense liquid phase by high pressure, expands and turns into a gas. As it vaporizes it absorbs the required latent heat of vaporization from the interior of the refrigerator. The inside of the appliance, because it gives up heat to the vaporizing Freon, becomes cooler.

The gasified Freon is then pumped back outside where it's subjected once again to high pressure and again liquified by the compressor. As it turns back into a liquid it releases the latent heat of vaporization it picked up inside the refrigerator. This heat is radiated outside the appliance to the surrounding environment and diffused. The reliquified Freon is pumped once more into the

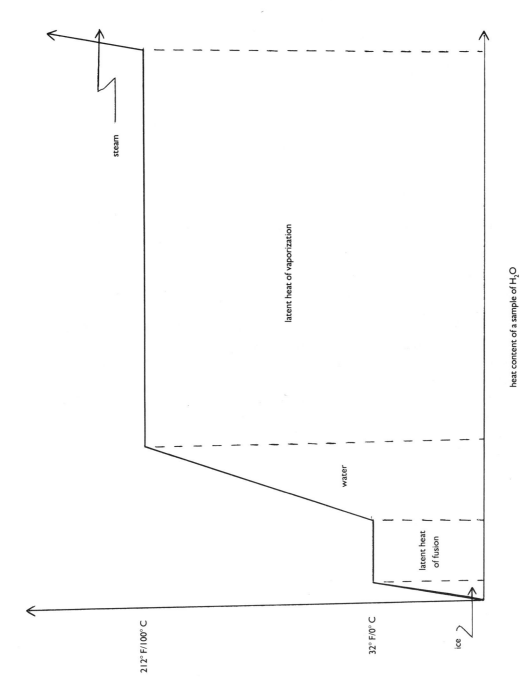

heat content of a sample of H₂O

Figure II.2   Heat vs. temperature

refrigerator and the cycle repeated until the desired temperature is achieved inside the refrigerator. Were it perfectly insulated, a refrigerator, once cooled, would stay cooled regardless of the type or quantity of its contents. Real, imperfect refrigerators must periodically work to remove the heat allowed in through less-than-perfect insulation, opened doors, and added contents.

Note that nowhere during the discussion of the refrigerator is there any reference made to anything like "coldness." A refrigerator doesn't really create cold. Nothing does. A refrigerator cools food only by removing heat from it. This heat isn't destroyed, merely redistributed. The Freon acts like a continual conveyor belt, hauling heat from inside the refrigerator and depositing it outside. It picks up heat as it vaporizes; it dumps it as it condenses. The heat that started out in the food ends up in the kitchen. This is why you can't cool your house by letting your refrigerator stand open. It won't eliminate any heat; it will only circulate it. To cool a house the heat absorbed indoors must be dumped outdoors. To accomplish this at least part of the cooling system, like the rear end of a window air conditioner, must lie outside the house.

The operation of a refrigerator serves to illustrate another important aspect of the transition between the liquid and gaseous states, namely the role played by pressure. The liquid Freon vaporizes when pressure is eliminated inside the appliance. The gaseous Freon is condensed by the application of pressure outside the appliance. The Freon is thus transformed from a liquid, into a gas, and back into a liquid again merely by altering the amount of pressure to which it's being subjected. The change of states here isn't a function of temperature, but of pressure.

Pressure has the ability to change a liquid to a gas and vice versa because of the drastic change in volume associated with condensation and vaporization. As a liquid vaporizes, its volume expands hundreds of times over. Pressure hinders this expansion and makes the gaseous state more difficult to attain. Even at temperatures that might otherwise allow vaporization, a high pressure will compress a substance enough to keep it in its liquid phase. A low pressure, on the other hand, facilitates the expansion involved in gasification. Low pressures encourage liquids to boil even at temperatures that otherwise wouldn't allow it.

One hundred degrees Celsius or 212 degrees Fahrenheit is the standard boiling point of water—the temperature at which it will boil under normal pressure, normal pressure being defined as one atmosphere of air at sea level. As pressure is reduced, however, so is water's boiling point. The less pressure it has to overcome as it expands into steam the more readily water will vaporize. Because the air pressure is lower at higher altitudes, water boils at a lower temperature at higher elevations. Because water boiling at a high altitude is cooler than water boiling at a low altitude it takes longer to cook something by boiling it in Denver than it does in Houston.

The only way to increase the temperature of a pot of boiling water in Denver is to increase the pressure at which it's being boiled. This can be

■■■■■■■■■■■■■■■■■■■■■■■■■■■■■■■■■

Melting Under Pressure

The role played by pressure as substances move back and forth between the solid and liquid states is similar to, although far less drastic than, that played at the liquid-gas frontier. Because nearly all substances expand slightly as they thaw, pressure generally encourages freezing and inhibits melting. Water, of course, is an important exception.

Because water expands as it freezes, high pressure tends to drive it into its relatively compact liquid state. It's the pressure of an ice skater's weight that melts the ice beneath her blades and creates the film of liquid water upon which she glides. Skating is difficult at extremely low temperatures because much less lubricating liquid is created.

The high pressure in an unopened soda bottle can keep the contents liquified even at subfreezing temperatures. When the cap is popped, of course, the pressure is released. The icy-cold soda, free to expand, quickly assumes a solid state. What starts out as a cold Dr. Pepper becomes, in a few seconds, a bottle of brown slush.

■■■■■■■■■■■■■■■■■■■■■■■■■■■■■■■■■

accomplished with the help of a device invented way back in 1679 by a French mechanic, Denis Papin (1647–1712). The so-called steam digester, now better known as a pressure cooker, traps created steam and thereby increases its internal pressure as it's heated. The increased pressure inhibits the boiling of water and allows water temperatures much higher than those found under normal conditions. Potatoes boiled in the superhot water within a pressure cooker cook hotter and faster than those cooked in an open pot.

II.13 | EVAPORATION

Water exists as a solid, a liquid, or a gas depending upon temperature. While this is generally true, it isn't absolutely true. Even at temperatures that favor water's solid or liquid phase at least a trace of water exists in the air as a vapor. Even when rivers, lakes, and swimming pools freeze there is some humidity.

Once again, this all makes perfect sense when you think of water as a mass of squirming molecules. It's the speed with which these molecules squirm that determines temperature and decides if water is going to be ice, liquid, or steam. But regardless of what sort of consensus the group reaches, there will always be those who refuse to comply.

This is because not all water molecules are equal when it comes to energy content. Some squirm with more vigor than others. Even at extremely cold temperatures a few exceptional individuals are able to wrench themselves away from their fellows. The escape of solitary molecules from solid water is called sublimation. A pile of snow will slowly shrink away to nothing without ever melting due to sublimation. Wash hung out to dry will eventually do so via sublimation even though the temperature never climbs above the freezing point.

Much more common and important than the sublimation of ice or snow is the transformation of liquid water to a gas. This is called evaporation. Evaporation is generally a much faster process, of course, because the group from which the refugees flee is at a much higher temperature. Indeed, the speed of evaporation depends a great deal upon the temperature at which it takes place. All other things being equal, hot water will evaporate more rapidly than cold water. Hair and clothes driers use great quantities of heat as part of their strategy to speed the evaporation process.

The rate of evaporation can also be enhanced by increasing the opportunities for a fast-moving molecule to make its break. This can be accomplished by making sure all molecules are reasonably close to the air. By spreading a given amount of water out over a larger area, the number of molecules at the surface is multiplied and their average distance from the surface is decreased. A glass of water has only one small, round area at the top through which molecules may escape. The same amount of water, splashed across a floor, has a much larger surface area and nearly all its molecules are positioned to make the break into the atmosphere. While the water contained in the glass might take a day and a half to evaporate, the same water, splashed across the floor, might take an hour and a half. Clothes are tumbled and hair tousled when dried in order to ensure that all surfaces are exposed to the air.

Low humidity also urges liquid water toward its gaseous state. In this case it's not so much that evaporation is accelerated, but rather that the reverse process, called condensation, is slowed down. Condensation happens when an already airborne molecule slams into a liquid and becomes trapped there. At a lower humidity there are fewer of these kamikazes darting about so less condensation occurs. With the condensation working against it minimized, evaporation enjoys greater success.

A breeze also cuts down on condensation and thereby enhances the net rate of evaporation. Moving air sweeps a molecular escapee away from the liquid it just left and reduces its chances of reentry. A pair of wet sneakers will dry more rapidly in front of a fan because the condensation retarding their drying is virtually eliminated. Hair and clothes driers rely as much on moving air as they do on hot air.

Regardless of the speed at which it takes place, vaporization via the relatively gentle process of evaporation requires the same amount of heat as does vaporization by boiling. Whenever evaporation takes place it requires the latent heat of vaporization. This rather considerable quantity of heat is drawn from the immediate surroundings, and those surroundings are cooled as they supply it. Blowing on a spoonful of hot soup speeds its evaporation, and thereby cools it.

A much more important application of the same principle takes place when we sweat. When we get hot we sweat. This sweat then evaporates. As it does so it takes its required latent heat of vaporization from our skin and so cools us. The swifter the net rate of evaporation the greater the cooling effect. A breeze, the removal of clothes, and low humidity all promote evaporation of sweat and our ability to cool ourselves. High humidity, clothing, and dead calm have the opposite effect.

On a large scale evaporation affects the very air we breathe. The enormous amount of water in the world's oceans, seas, and lakes is perpetually being turned into water vapor via evaporation. Aided by huge surface areas, prevailing winds, and the energizing rays of the sun, thousands of tons of water enter the air every second of every day. This process simultaneously cools the earth's surface and creates a vast amount of humidity.

II.14 | WEATHER EVENTS

The steady movement of water from its liquid to its gaseous state via evaporation, along with the formation of liquid water from water vapor via condensation, defines many of our weather events. The starting point for understanding how they work is humidity.

Humidity, simply defined, is the presence of gaseous molecules of water in the air. As you have probably noticed, humidity is generally discussed and dealt with as relative humidity. Whereas absolute humidity measures the quantity of water vapor in a sample of air, relative humidity compares the actual water content of the air to its theoretical maximum content. Humidity is normally calculated in relative terms because the measure of relative humidity provides the best gauge of water's ability to soak up more water and dictate the weather.

Relative humidity is expressed as a percentage. If the relative humidity is 5 percent it means that the air contains only 5 percent of the molecules of water it could hold if it were fully saturated. Air this dry, generally found only in extreme desert conditions, has a tremendous potential for absorbing water and does so readily. Anything moist—soil, plants, animals, exposed skin, and lips—will dry out quickly when the relative humidity is this low.

A relative humidity of 50 percent means that the air is half as humid as it could be, or halfway saturated. Halfway or partially saturated air is commonly encountered under many weather conditions and is desirable for human comfort. Evaporation takes place, but not to any extreme degree. Laundry will dry, but lips won't crack.

A relative humidity of 100 percent means the air is fully saturated and incapable of absorbing any more water vapor. At 100 percent relative humidity evaporation and condensation are in a state of equilibrium. For every molecule of water that joins the atmosphere another one drops out. The result is a net evaporation rate of zero. Saturated or nearly saturated air can be experienced during or immediately after a summer shower or on a dreary, drizzly sort of day. This much humidity is often regarded as unhealthy and always considered uncomfortable. It has the unfortunate ability to make hot weather seem hotter and cold weather seem colder.

The difference between absolute and relative humidity arises because air's ability to hold moisture changes dramatically depending upon its temperature. Air at water's freezing point, for example, can hold about 4 grams of water per cubic meter. Room-temperature air, meanwhile, can hold about 16 grams of water per cubic meter. Freezing-point air with a content of 4 grams of water per cubic meter and room-temperature air with a content of 16 grams of water per cubic meter are both saturated and are both at 100 percent relative humidity. Neither is any good for evaporation purposes even though the absolute humidity in the freezing air is only one-fourth as great as that of the room-temperature air.

Let's take our saturated, freezing air and warm it to room temperature. After warming, its absolute water content remains the same—it still has 4 grams of water per cubic meter. But its relative humidity has dropped from 100 percent to 25 percent as now, instead of being saturated, it holds only a fourth of the water it's capable of holding. When cooled to the freezing point again its relative humidity will return to 100 percent. Other things remaining equal, an increase in temperature results in a decrease in relative humidity and a decrease in temperature results in an increase in relative humidity.

It's the warmest air that has the greatest capacity to hold water and the most power to absorb it. On a global scale, the equatorial atmosphere drinks up ocean water like a giant sponge. This warm, now-moist air, being less dense than cool, dry air, tends to rise. (See Unit I.14.) Denser air flows in from the north and south to take its place, setting up a prevailing system of winds as it does so. The moisture-laden air is circulated to the north and south. As it wanders farther from the equator it cools. Its relative humidity increases and eventually reaches 100 percent. When this saturated air is cooled even further it relinquishes the moisture it can no longer hold. The excess moisture condenses in midair as tiny droplets or, if the temperature is very cold, as tiny ice crystals. The resultant condensation masses together and becomes visible as a cloud.

Once tiny droplets or crystals are formed they begin to coalesce into larger and larger assemblages. Eventually these conglomerations of liquid or solid water become so heavy that they no longer remain suspended as a cloud, but fall to earth. The liquid drops, or the frozen crystals that melt as they fall, come down as rain. Frozen crystals that remain frozen as they fall come down as snow. Sometimes the liquid drops freeze as they fall. This results in sleet. Under special conditions sleet may form, then be carried back into a cloud by an updraft, fall again, be blown back up again, and so on in a repeating cycle. When this happens the sleet particles acquire a new layer of ice during each

■■■■■■■■■■■■■■■■■■■■■■■■■■■■■■■■■■■■■

Condensation Nuclei

In order for humidity in the air to condense into tiny water droplets it generally needs something upon or around which to condense. Dust, smoke, and other impurities in the air usually provide these so-called condensation nuclei. Without them, the droplets may fail to form and the air can become supersaturated, holding onto unnatural amounts of water vapor.

In an effort to speed the condensation process and induce rain, clouds are sometimes seeded with impurities. These usually take the form of dry-ice crystals that not only provide condensation nuclei but also lower the temperature of the air and thus increase relative humidity.

Supersaturation is most common at high altitudes where the air is very clean. Jets that travel through this supersaturated air provide a stream of condensation nuclei in the form of exhaust particles. These initiate the condensation process that results in the formation of the long, skinny clouds called contrails.

■■■■■■■■■■■■■■■■■■■■■■■■■■■■■■■■■■■■■

round-trip and grow like pearls inside an oyster. These multilayered balls, once they gain enough weight to finally tumble to the ground, fall as hail. Their final size is determined by how many times they've been tossed about in the sky.

Condensation can also occur if surface air is suddenly cooled. Fog is basically the formation of a cloud at zero altitude. This often takes place as wet air drifts in off a warm body of water and onto a cooler landmass. When this happens the fog can be seen "rolling in." Fog also forms during the cooling that takes place at nightfall and generally continues on into the first hours of daylight. On a smaller scale, the same thing happens to your breath on a cold day. The warm, moist air from your lungs cools as it hits the frigid outside air. Its capacity to hold moisture diminishes and the water vapor it held condenses into tiny, visible droplets.

A more common overnight event than fog is the formation of dew. In this instance only the air in direct contact with the ground is cooled to what is called the dew point. At the dew point the air is cooled to where it can no longer hold onto its humidity. It gives up its excess moisture as dew. The exact same thing happens on a summer's day when hot, wet air is cooled as it comes into contact with a toilet tank or a chilled drink. Dew is generally heaviest on clear nights when the earth's surface can most effectively radiate away its daytime warmth, and on still nights when the contact between ground and air is least disturbed.

The dew point is nothing more than the temperature at which relative humidity achieves the 100 percent level. On very humid days the dew point may be only a few degrees below daytime temperatures. On dry days a great deal of chilling is needed to drop the temperature to where the air would be saturated. Sometimes the dew point falls below the freezing point. Under these conditions frost, not dew, will be formed. During the formation of frost, water goes directly from its gaseous to its solid phase, building intricate crystal latticeworks molecule by molecule as it does so.

II.15 THE FLOW OF HEAT

Whether it's involved in the operation of a refrigerator or the creation of a weather system, heat constantly tries to equalize its presence. Heat always flows from where it's relatively highly concentrated to where it's relatively dilute. Heat always flows from where the temperature is high to where it's low. Heat never flows in the opposite direction, nor does it ever fail to flow in the direction of cold when given the opportunity. The irreversible and irresistible flow of heat from what is hot to what is cold constitutes the second law of thermodynamics. (The first law of thermodynamics is the law of the conservation of energy, discussed in Unit II.2.)

The transfer of heat from high to low concentrations proceeds in three ways: conduction, convection, and radiation. (See Figure II.3.) Conduction is the process whereby heat spreads from one part of an object to the rest of the object, or from one object to another object with which it has direct contact. Conduction transfers heat from an electric stove's burner to the bottom of a frying pan and from the pan's bottom to an egg being fried. The egg never touches the burner, but gets hot just the same. The heat it receives has been transferred from the burner through an unbroken chain of conduction.

Some materials, appropriately called conductors, conduct heat rapidly and efficiently. Others, called insulators, conduct heat slowly and poorly. Most metals, like aluminum and cast iron, are good conductors. Wood, rubber, and plastic are good insulators. An aluminum ice-cube tray in one hand and a plastic one in the other hand, both taken from the same freezer, are at the same temperature. The aluminum one feels colder, though, because it conducts more heat away from the hand holding it than does the plastic one.

The second method of heat transfer, convection, involves indirect rather than direct contact between the objects involved. The heat moves not from a hot object to a cold one, but from a hot object to a fluid intermediary, and then from this intermediary to the second object. The intermediary agent is usually air or water. Whereas frying operates by conduction, baking operates by convection. Heat is transferred from the oven's heating element to the bread by means of the intermediate air. Most furnaces work by means of convection. The heat generated in the furnace is used to warm air, and this air then circulates throughout the house warming its contents and occupants. The coolant in an automobile's cooling system also transfers heat by means of

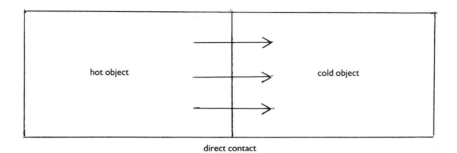

heat transfer via conduction

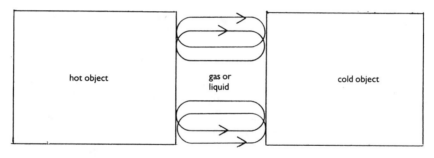

heat transfer via convection

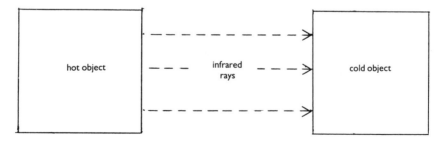

heat transfer via radiation

Figure II.3   Three methods of heat transfer

convection. The heat from the engine is transferred to the coolant, the coolant circulates through the radiator, and the heat is there transferred to the air. On grander scales the winds and ocean currents carry enormous amounts of heat from place to place through air and water intermediaries.

The transfer of heat by convection can be retarded by either eliminating the fluid intermediary or by halting its circulation. A car overheats when the radiator runs dry because there's no longer any fluid present to transport the heat away from the engine via convection. A car also overheats when the water pump goes out because the coolant is no longer forced along its path from engine to radiator.

The third form of heat transference is called radiation. When heat is transported by means of radiation, the energy involved temporarily loses its identity as heat and exists as a form of radiant energy called electromagnetic radiation during the course of its journey. (See Unit II.25.) Heat is transformed into radiant energy at point A, usually by something that is hot enough to glow. The radiant energy then travels from point A to point B and, once at point B, becomes heat once again. The transfer of heat by radiant means requires neither direct nor indirect contact among the objects involved. Broiling is a good everyday example of radiant heat transfer. The heat from the red-hot broiler gets to the meat neither by touch nor by circulating air but solely by radiation.

Radiant energy, as well as being able to travel though a vacuum, can also penetrate some materials, especially transparent ones. The sun's rays not only pass through empty space, but through the atmosphere and then the windshield of a parked car before they strike its interior. When the radiant energy strikes the seat covers and dash, it becomes heat. The car's interior, unable to generate radiant energy, can get rid of the heat only by convection or conduction. If the windows are closed convection will be almost nil. Convection or conduction work much more slowly than does radiation, so the outgoing flow of heat is much less than the incoming one. Heat accumulates inside the car and makes it very hot. A greenhouse works the same way, as do many types of solar-powered heaters.

To prevent the flow of radiant heat to areas where it's unwanted it's necessary to either intercept it or deflect it. Interception is easily accomplished by placing any opaque obstacle between the source of the radiation and the object or area to be protected. Sun hats, umbrellas, elm trees, and anything else that creates shade are frequently employed in this capacity.

Deflection, or reflection, is best accomplished with objects that are opaque as well as either light-colored or shiny. A black car interior reflects almost no radiant heat, but rather absorbs it. A white interior, in contrast, will bounce some radiant energy right back out through the windshield without ever absorbing it. Light-colored clothing is more comfortable in the summer for the same reason. (See Unit II.24.)

Stopping Heat

Imagine you are a thermodynamics engineer. Your assignment is to build a container that keeps hot liquids hot or cold ones cold as long as possible within the limits imposed by the second law of thermodynamics.

Keeping hot liquids hot means retarding the flow of heat from inside your container to the relatively cool outside. Keeping cold liquids cold, on the other hand, means retarding the flow of heat from the outside in. In either case, you need to minimize as much as possible the transfer of heat by each of the three means it travels: conduction, convection, and radiation.

In order to combat conduction you know you have to build your container out of a material with a high insulation value. Based on texts and charts at your disposal you review the possibilities: wood, rubber, styrofoam. The best of them all, however, is air.

Building a container out of air is clearly impossible, but you cleverly devise a way to incorporate a thin blanket of air into your design. This you do by fashioning two bottles of different sizes and suspending the smaller one inside the larger one.

You next need to minimize the effects of convection. You decide at first to seal the bottles at the neck and thereby halt any circulation of the trapped air. Then you realize that before you make the seal you can suck the air out from between the two bottles and virtually eliminate heat transfer via convection. There's a bonus here as well, because a vacuum is an even better shield than air against the flow of heat by conduction.

Finally, you need to do what you can to prevent the radiant flow of heat. Because you want your container to act against the flow of heat in both directions, you make both the outer surface of the inner bottle and the inner surface of the outer bottle highly reflective.

You call your invention a vacuum bottle, but it becomes better known by its trade name, Thermos®.

—Which methods of heat transfer are retarded by tinted windows? Double-pane windows?

—How do clothes help keep us warm?

—Why does wind make a cold day feel colder? Does the wind affect the reading on a thermometer?

—Why does a swimming pool feel so cold even though it's at the same temperature as the air, which is comfortable?

■■■■■■■■■■■■■■■■■■■■■■■■■■■■■■■■■■■

The Greenhouse Effect

The earth is, in some regards, like a giant greenhouse. The atmosphere is the glass in this analogy. It allows the radiant energy of the sun to reach the planet, then acts as an insulator that prevents the loss of heat via convection or conduction, thus maintaining the globe at a stable and appropriate temperature for its inhabitants.

There is now concern that escalating pollution of the atmosphere will impair the operation of this delicate mechanism and cause widespread, harmful, and irreversible disruption of the world's climate. According to the most popular theory, the buildup of carbon dioxide and other invisible pollutants will actually enhance the greenhouse effect and thereby cause global overheating. An alternate fear is that the continual accumulation of dust, smoke, and other particulate pollution will eventually reduce the amount of energy we receive from the sun and bring on another ice age.

Some say fire, some say ice. There's evidence to support both theories, but which one, if either, will prevail over time is anybody's guess.

■■■■■■■■■■■■■■■■■■■■■■■■■■■■■■■■■■■

II.16 SOUND

When vast conglomerations of molecules, identifiable as physical objects, move in unison their motion is detected as mechanical energy and described in terms of Newton's three laws of motion. When hordes of individual molecules vibrate randomly on a submicroscopic scale their motion is detected as heat and described in terms of the laws of thermodynamics. There's a third type of energy, containing elements of each of these but distinct from either, that also depends directly upon the phenomenon of movement. This form of energy is described according to the laws of acoustics and is detected as sound.

The type of motion that produces sound is neither the massive, bulky movement of mechanics nor the tiny, fragmented quiverings of heat, but rather a hybrid of the two. It ultimately involves molecules, of course, and in vast numbers. But when molecules move to produce sound they do so not as a single, welded lump as they do in mechanics nor as a hodgepodge of independent individuals as they do in thermodynamics. Sound results when populations of molecules dance in a coordinated and rhythmic manner. The ebbing and flowing is neither perceptibly large as it is in mechanics nor immeasurably small as it is in thermodynamics, but somewhere in between.

Sound, like any other form of energy, can be created from or converted into different types of energy according to the law of the conservation of energy. The mechanical blast of air through a set of vocal cords produces sound. The sound thus produced can be converted into electrical energy by a telephone. After being transported over long distances in the form of electrical impulses, the electrical energy can once again be converted into sound by another telephone.

When not intentionally captured or transformed into other forms of energy, sound generally disperses and eventually ends up as minute amounts of heat. The organized pulsing of sound energy inevitably deteriorates into the diffuse, undisciplined molecular mayhem of thermal energy. Sound isn't often a very powerful energy form, and the heat produced from it is rarely significant. Still, when loud enough, sound can possess a good deal of energy. Thunder and sonic booms are capable of shattering windows and shaking buildings.

Sound is loud or soft, shrill or muted, pleasant or irritating, absorbed, conducted, or reflected depending upon the action of the energy patterns by which it's defined and characterized. These patterns are called waves and, to a certain extent, resemble the more familiar waves generated when a stone is tossed into the middle of a still pond.

A sound source sends out expanding waves of sound energy the same way the stone sends out ever-growing circles of ripples. In the case of the stone, the energy is transported via the fluid water. In the case of sound, the conducting medium is nearly always air. In each case, as the waves travel farther from their source they spread out evenly in all directions and at a constant speed. At the same time they're continually weakened as their energy content is diluted over an ever-greater area. Eventually they become so weak that they lose their identity as waves.

Although sound waves may superficially resemble water waves, they're fundamentally quite different. Water waves are transverse waves while sound waves are longitudinal waves. In a transverse wave the motion of the conducting medium is perpendicular to that of the energy being transmitted. As a ripple moves horizontally across a pond the water at the pond's surface moves vertically. In a longitudinal wave the conducting medium moves in the same direction as does the transmitted energy. The air molecules conducting the blare of a teenager's stereo to her ear oscillate to and fro along the same path the sound follows. No individual molecule travels very far, however. Each jumps toward the ear then recoils to its original position after bouncing off its nearest neighbor down the line. The molecules behave something like the firemen in a bucket brigade. The sound energy, like the buckets, is passed along from one molecule to the next while each stays more or less in place.

As sound waves travel they create alternating areas of colliding and rebounding molecules. These correspond to the crests and troughs of familiar water waves. The areas where collisions are occurring contain densely packed molecules and are referred to as either compressions or condensations. Where molecules are rebounding they're more spread out, forming what are variously called expansions or rarefactions. One complete sound wave consists of one compression and one expansion. A wave that carries more energy produces more violent collisions and creates more pronounced condensations and rarefactions. A higher-energy wave conducts a sound of greater intensity and is perceived as being louder than a lower-energy wave. A wave that has more rapidly alternating condensations and rarefactions results in a sound of higher frequency. High-frequency waves are perceived as being of higher pitch than low-frequency waves.

II.17 ACOUSTICS

Any sound, regardless of its volume or pitch, depends upon two or possibly three things in order to exist. The first thing a sound needs is a source to generate it, the second thing is a medium of transmission to carry it, and the third, optional, thing is an ear to hear it. In a purely physical sense sound exists when only the first two factors are present. These by themselves are enough to set up and sustain the form of energy known as sound and studied as acoustics. This energy form exists whether or not an ear is present to hear it, and a tree falling in a distant forest falls with a loud crash under this definition. In a physiological sense, though, sound depends upon a receptive ear. According to this definition the tree falls silently.

Physicists and physiologists agree that the source of sound is a mechanical movement of some sort. Just as a stone is needed to set the surface of a pond into motion, so a physical object, appropriately vibrating, is needed to set sound waves on their way. A plucked violin string, water crashing on a beach, an exploding shotgun shell, and a fingernail being dragged across a blackboard all vibrate in ways capable of producing sound. It's often possible to sense these vibrations by other means than the ear. A violin or guitar string can be seen to move. The movement of a ringing bell or a rattling license plate can be felt with the fingertips. In any event it's the frequency, intensity, and complexity of the vibration that determines whether the resultant sound will be a concerto or a cacophony.

Besides vibrating in an appropriate manner, each object, in order to produce its distinctive sound, must also vibrate in the presence of something that will sense its mechanical movement, translate it into sound waves, and propagate it. A stone tossed into a dried-up pond makes no waves, and a tree falling in an airless forest would make no sound even in the presence of many ears. Robert Boyle discovered this fact in 1660 when he suspended a ringing alarm clock under a glass dome. As he pumped air out from under the dome the alarm became fainter and fainter until it finally disappeared altogether. The vacuum Boyle created, being void of any material medium, was incapable of picking up the vibration of the clock's bell and transporting it as sound.

In general, a substance's ability to conduct sound is directly related to its elasticity—its tendency to resume its original position and shape after being disturbed. This holds equally true for solids, liquids, and gases. Some solids, like wood, cork, clay, or cotton, are very inelastic. When subjected to stress they absorb energy rather than transmitting it. These inelastic substances are poor conductors of sound or good sound insulators, depending upon how you look at it. Other solids, like steel, resist stress, passing it along rather than absorbing it. These highly elastic substances are good conductors of sound. By putting your ear to a railroad track you can hear a train that's far out of sight. And lying in bed you can hear distant toilets being flushed as the sound is carried throughout your apartment building by the rigid network of plumbing.

Most liquids possess a similar degree of elasticity—one that lies somewhere between the inelasticity of cork and the high elasticity of steel. Their ability to transport sound is thus also rather similar and lies between these extremes. Anyone who has ever had their head under water knows that sounds travel through liquids fairly well. Not as well as through steel, perhaps, but better than through wood or clay.

Far and away the most common carrier of sound is neither a solid nor a liquid, but that most familiar of all gases—air. Air can carry sound because it's a physical substance and because it is, to a degree, elastic. Air subjected to the shimmy of a saxophone reed, the percussion of a slamming door, or the shudder of a bouncing ball will be jolted back and forth by these mechanical motions in such a way as to form sound waves. Being less elastic than most solids and all liquids, air is really a rather poor conductor of sound. Nevertheless, because our ears are normally immersed in a sea of the stuff, connecting us to everything we hear, air is considered the standard means by which sound travels.

Left alone to travel through an uninterrupted expanse of air, sound waves eventually tire, become weaker, and deteriorate into the random motions of heat. Should they first strike a soft surface like a carpet they meet the same fate, only a bit sooner. The difference is that in the latter case it's the carpet, and not the air itself, that absorbs the acoustical energy and transforms it into minute amounts of heat. If sound waves strike hard but uneven surfaces, like those in a roomful of furniture, their energy will bounce off those surfaces, be scattered incoherently, and again become heat in one way or another.

If a sound wave strikes a relatively smooth and hard surface like a tile floor, though, it will be reflected more or less intact. It will retain its identity as sound and continue to travel, although in an altered direction. The reflected wave will closely resemble the original wave, but will be somewhat reduced in intensity. The reflected wave is then capable of striking a soft surface or a hard but uneven surface and ending up as heat. But it's also capable of striking another smooth and hard surface and being reflected again, producing yet another similar but weaker wave, which can again be reflected, and so on. This process will continue until the repeated reflections sap the wave of all its energy. When this repetition of reflections takes place within a confined area, like a shower stall or high-school gymnasium say, the reflected waves prolong the perception of sound. This effect is called resonance or reverberation.

Reverberation can be good or bad. Acoustical studies have determined that a one- to two-second period of reverberation makes for a lively but still clean sound. Churches, lecture halls, and auditoriums are often designed to produce such an effect. The speed of sound, the surfaces upon which it lands, and the amount and angles of reflection are all taken into account in this endeavor. Shorter reverberation periods result in a dead sound, longer periods in a blurred sound. Sometimes a very long reverberation period contains a period of silence between the passage of the primary wave and its reflection. The secondary wave in this case is heard not as part of the first wave, but as a separate sound. Such a sound is called an echo. When you shout across a canyon you send out a pattern of sound waves. These waves strike the far wall, bounce off it, and return to you in pretty much the same condition they were in when they left your mouth. The time between your original shout and the perception of the echo is how long it took the energy to traverse the width of the canyon to and from the opposite wall.

II.18 THE SPEED OF SOUND

Air is such a common medium for sound that it's usually taken for granted. In the absence of any disclaimers, a sound wave is assumed to be a sound wave traveling through air. And the speed of sound, unless specifically stated otherwise, is assumed to be the speed of sound through air. In fact, though, sound has a multitude of different speeds depending upon what it's traveling through. Its speed, like the distance it travels, is directly related to the elasticity of its means of transport. Sound moves through steel about four times as fast as it does through water, and through water about four times as fast as through air. Its speed through air, though, is of particular importance for both popular and practical reasons.

The speed of sound through air was first calculated reliably in 1640 by a French mathematician, Marin Mersenne (1588–1648). He measured the time it took his voice to cover a measured distance to a wall and then back to his ears as an echo. The resulting speed of 343 meters per second was an item of intellectual curiosity at first, but has since become an important scientific and engineering standard. The acoustical design of buildings, musical instruments, mufflers, and loudspeaker systems can only be accomplished by calculating the movement of sound waves at known velocities.

In many regards the speed of sound is really quite slow. The fact that it can dawdle long enough getting to and from the far side of a canyon to cause an echo attests to that. The perceptibly slow speed of sound can actually be quite a nuisance. The message of a graduating valedictorian may be all but incomprehensible because of the long reverberation time in the gym where she's giving her speech. At large outdoor concerts the sound of a lead guitar solo may actually take a second or more to reach the farthest fringes of the crowd. Distant auxiliary speakers have to be carefully synchronized to reinforce rather than interfere with the sound originating from the stage.

In other regards the speed of sound is exceptionally fast. For many years it was unimaginable that any physical object, much less a human being, would ever travel at such a velocity. But the advancement of aerodynamics and jet propulsion eventually made it a possibility. Even after the technology was developed, though, there remained a great deal of heated debate about whether or not it should be attempted. Traveling faster than the speed of sound marked a step into the unknown and was met by all the anxiety, fear, and superstition that normally accompanies such events. All these reservations were swept aside in October of 1947, however, when Charles E. Yeager (1923–) flew the experimental rocket-powered X-1 aircraft faster than the speed of sound. He was the first human to achieve supersonic flight and thereby break what is variously called the sonic barrier or the sound barrier.

The speed of sound isn't just an abstract idea or a mathematical abstraction. Unlike the highway speed limit, the sound barrier is a real physical obstacle that is a function of the elasticity of air, air's capacity to absorb and then rebound from any sort of disturbance. A sound wave is a disturbance and 343 meters per second is the speed at which air molecules bounce back and forth among themselves as they participate in the chain-reaction collisions that transport sound waves. The speed of sound, then, is determined by the natural agility of air molecules.

The dexterity of air molecules allows them to bounce off of and get out of the way of anything traveling at subsonic speeds, anything moving slower than 343 meters per second. Once something travels at speeds greater that this, air molecules are no longer capable of evasive action. A plane traveling at a subsonic speed allows air molecules to bounce off it and get out of its way. A plane traveling at a supersonic speed doesn't. The air in front of a supersonic plane is trapped against it and piles up in front of it like snow before a plow. The squeezed air molecules form a dense wall of swirling, bubbling air that the plane must penetrate. Only specially designed planes can withstand the strain that accompanies this effort. As the wall of air being pushed along in front of the plane collides with still air an area of tremendous compression is formed. This compression, along with its associated rarefactions, constitutes an extraordinary set of sound waves called a sonic boom.

Anything—not just an airplane—traveling at supersonic speed will cause a similar shock wave and produce a similar boom. A bolt of lightning heats the air it passes through so quickly that the air expands at a rate faster than the speed of sound. The supersonic shock wave is heard as thunder. A rapid combustion of fuel, called an explosion, produces the same effect. And the crack of a whip is the result of its tip moving at a supersonic speed.

● ●

Mach 1

Ernst Mach was born in Austria in 1838 and enjoyed considerable success there both as a scientist and as a philosopher. It was during the scientific phase of his career at the University of Prague that he performed his studies on the behavior of objects as they moved through air at high speeds.

It didn't take Mach long to notice that the speed of sound was of special importance in his experiments. The behavior of objects changed drastically depending upon whether they traveled below, at, or above the speed of sound.

Although seemingly straightforward, this finding was complicated by the fact that the speed of sound, even through air, is far from constant. Humid air conducts sound a bit faster than does dry air. Warm air transports sound quite a bit faster than cold air does. This holds true for any sound, regardless of its volume, pitch, source, or direction.

What really interested Mach, then, wasn't an object's absolute speed in terms of miles per hour or meters per second, but its speed relative to that of sound under the same conditions. To this end he devised a scale to deal with the ever-changing speed of sound. The speed of an object traveling at the local speed of sound, whatever it might happen to be, measured 1.0 on his scale. An object traveling half as fast measured 0.5, twice as fast 2.0, and so forth.

Although this scale made Mach's work easier, it was, at the time, of no great interest to the scientific community in general. When he moved on to other interests in 1895 it was all but forgotten. It remained that way until 50 years later when airplanes began to approach the speed of sound.

Now, with the shattering of the sound barrier an everyday occurrence, Mach's name has become a household word. The speed of sound under local conditions is universally known as mach 1. A plane flying at a slower speed, at any mach number lower than 1, is said to be subsonic. A plane flying at a higher speed, at a mach number greater than 1, is said to be supersonic.

—Is the mach scale useful in outer space? Why or why not?

—Is it easier to break the sound barrier at sea level or at high altitudes?

—Can the sonic boom created by supersonic flight be heard from inside the plane?

—Why is supersonic flight often prohibited over populated areas?

● ●

II.19 THE SUBTLETIES OF HEARING

While sound as a form of energy behaves strictly according to the physical laws of acoustics, its impact upon us as human beings is purely a physiological phenomenon. Sound energy is merely the raw material for the process of hearing. The ear, along with its associated components in the nervous system, is the apparatus that detects the molecular ebbings and flowings of the air, interprets them as meaningful patterns, and assigns to them all sorts of emotional, psychological, and intellectual values.

The human ear is capable of hearing sound waves with a frequency of anywhere from 20 to 20,000 cycles per second. This allows us to hear anything within the 85- to 1,100-cycle range the human voice is capable of making as well as quite a bit more—everything from the rumble of distant thunder to the whine of a high-speed drill. Still, our abilities in this area are somewhat limited. There exist all sorts of waves that lie outside the range of human hearing.

Physiologically speaking, of course, a wave with a frequency of below 20 cycles per second or above 20,000 cycles per second isn't really sound as we humans know it, because it can't be heard. Physically speaking, though, such waves are a very real part of the natural world. Infrasound, consisting of extremely low-frequency waves, can be detected by some animals as well as by the human-built devices designed to do so. Earthquakes produce these sorts of waves, and the ability of animals to "predict" them is based on their ability to hear what we can't. Animals also hear a variety of ultrasound waves with frequencies as high as 120,000 cycles per second. Dogs respond to "silent" whistles and bats navigate by emitting and listening to the echoes of a constant stream of "sound" unheard by humans. Although we can't experience these ultrasound waves directly, we're well aware of their existence and have put them to work knocking dust particles out of the air, killing bacteria, driving away cockroaches, pasteurizing milk, and drilling teeth.

The human ear's capability regarding intensity is a bit more impressive, although it still pales when compared to that of certain other species. Nonetheless, we can hear anything from a whisper across an empty theater to an F-16 accelerating for takeoff. Measured objectively, the loudest sounds we can hear are many billions of times more intense than the faintest. Sound waves with intensities too weak for the human ear to detect are said to lie below the threshold of hearing, while those containing more energy than the ear can cope with lie beyond the threshold of pain and do, in fact, cause pain, along with damage to the auditory system.

The abilities of the human ear, even within its somewhat limited range of intensity and frequency, are nothing short of amazing. The subtleties of hearing allow us to accurately differentiate among millions of different sound patterns and perceive minute variations among them. We can pick the voice of a loved one out of a crowd and tell if that loved one has a cold, is depressed, or lying.

■■■■■■■■■■■■■■■■■■■■■■■■■■■■■■■■■■■■■

Intensity and Loudness

Sound intensity is measured in units called decibels (db). The decibel scale is designed so that the faintest audible sound registers as zero decibels and the loudest, at the threshold of pain, 130 db. A whisper rates about 20 db, a normal speaking voice 60 db, and a rock band 120 db or more. Because of the way the scale is constructed a normal speaking voice actually contains 10,000 times as much energy as a whisper and a rock band a million times more energy than that.

It should be noted that decibels measure the energy content of sound in a purely objective manner. What we subjectively experience as loudness is a related but somewhat different thing. Any attempt to quantify the physiological impact of sound, in fact, is doomed to failure. Federal law, for example, limits the intensity of commercials on television and radio to the same level as that of the ambient programming. But advertisers, ever resourceful, have been able to devise ways of enhancing the perceived loudness of their messages without violating the technical intensity limits.

■■■■■■■■■■■■■■■■■■■■■■■■■■■■■■■■■■■■

Besides being able to identify a bewildering variety of sounds, we can also tell where they're coming from with an uncanny ability. This is because we have two ears. In fact, it's likely that every species of hearing animal has evolved with a pair of ears rather than only one in order to use them in tandem to locate prey or predators, as the case may be. Because the ears in any pair are separated by a distance of approximately one head, each ear is located slightly differently in relationship to any sound source and receives a slightly different pattern of sound waves. These differences from one ear to the other are of three distinct types.

To begin with, one ear is usually a bit closer or farther away from any given sound source than is the other. Incoming sound therefore reaches one ear slightly before the other. The hearing mechanism can distinguish this time lag from ear to ear and uses it to judge the direction from which a sound is coming. Also, because sound waves consist of an alternating pattern of condensations and expansions, each ear is located within a slightly different portion of any passing wave. One ear may be experiencing a compression while the other is in the midst of an expansion. Again, the nervous system is such that it can judge these differences, gauge them as they change, and use them to estimate the direction from which a sound is coming.

The presence of the head also comes into play. More than simply holding the ears apart, it puts between them a rather solid object that effectively absorbs and deflects incident sound waves. An ear directly in line with an incoming signal receives it more strongly than does the far ear, which is shielded from the sound by a dense mass of bone and tissue.

The human ear, besides being able to identify and locate a sound, can also tell if it's coming or going. Sound from an approaching source will grow steadily louder and from a receding one steadily fainter. But aside from these obvious clues, there's also something called the Doppler effect that provides us with further information.

First explained by Christian Johann Doppler (1803–1853), an Austrian physicist, in 1842, the Doppler effect is most noticeable when listening to the siren of a passing fire truck. The sound of the siren in this case gradually increases not only in intensity but also in frequency as it approaches, then drops in both regards as it passes. As the distance from source to listener shrinks the incoming sound waves are crowded together, the frequency of their arrival increased, and their pitch accordingly raised. After the fire truck passes the effect is reversed. The waves are strung out over time and distance and are heard at a lower pitch. The ear, familiar with the Doppler effect, correctly interprets the motion of the passing fire truck.

II.20 | VISION

Our ears provide us with a valuable means of detecting, understanding, and responding to the physical world. They alert us to the presence of anything vibrating within a certain range of frequencies by reacting to the sound energy it emits. By the sense of hearing, a vibrating physical object can be identified, located, and tracked with impressive accuracy. Hearing is a basic survival skill and arguably our most important link to the physical world. If it's not, then the sense of sight surely must be.

The sense of sight has more in common with the sense of hearing than importance. Like hearing, vision is a one-way channel of communication that receives but doesn't transmit. Both senses rely on a pair of receptors mounted high off the ground where they can gather the most input, and far enough apart to provide multiple vantage points. In each case the input gathered is a form of energy. The energy sensed by the ears is sound. That sensed by the eyes is light.

Light, like sound and every other energy form, behaves in accordance with the law of the conservation of energy. It can be created from and transformed into other types of energy but can't be conjured up out of nothing or absolutely destroyed. Light most frequently arises from chemical energy. When the chemical constituents of a candle react with atmospheric oxygen they generate a certain amount of light. Combustion, in fact, is defined as any chemical reaction that produces light. Light can also be produced by mechanical means. The glow of sparks flying off a grinding wheel was mechanical energy a moment earlier. A light bulb creates light from electrical energy, as does a red-hot toaster or electric burner. Each of these is an example of incandescence—the process that turns other forms of energy into light as an accompaniment to heat.

Incandescence is the most common but by no means the only light-yielding transformation of energy. Light can also be produced without heat. Such a process is called fluorescence. A fluorescent light bulb produces light by a chemical process that generates far less heat by-product than does its incandescent counterpart. A firefly generates fluorescent light by a biological process that can ultimately be explained in chemical terms. But whether it generates light by means of incandescence or fluorescence, any light-producing object or event is said to be luminous.

Light turns into other forms of energy by a similar variety of methods. Light falling on a living plant is converted into chemical energy by photosynthesis. Light can also be converted into electrical energy by what's called the photoelectric effect. Solar-powered instruments like calculators rely on this phenomenon to generate the electricity they need to operate. Television cameras are sophisticated photoelectric devices that pick up patterns of light and convert them into electrical signals. The signals are then converted back into patterns of light by a television set, which itself is a sort of sophisticated fluorescent light bulb. But most light, like most sound, eventually loses its energy to the medium through which it passes or to the surfaces it strikes and there ends up as scattered, minute bits of heat. Perhaps most interesting, though, is that tiny portion of light that enters an eye and ends up stimulating a nervous system.

Light, in the final analysis, is all we really see. We don't see waterfalls, rainbows, cousins, potholes, memos, or hamburgers. We see the light rays that radiate from them and end up in our eyes. In the case of luminous objects we see light that has originated at that object and from there traveled directly to us. Campfires, brake lights, lightning bolts, stars, and fireflies generate their own light, and it's that very same light that triggers the vision process. Everything else, all objects that don't glow, are made known to us by means of reflected light. Firewood, fenders, clouds, the moon, and mosquitoes, being nonluminous, need to be illuminated by some other light source before they can be detected by the eye. In the absence of an external light source they all remain invisible.

Because light travels along straight paths the source of any ray that strikes an eye, be it an original luminous source or a secondary illuminated source, can be located simply by sighting down the ray. In this way, barring any optical illusions, we see things as they are and in the direction they truly lie. It's due to the straight trajectory of light rays that we can reach out and pick up a telephone, guide a car down the highway, cut an onion, or walk through a doorway without colliding with the wall.

To precisely locate an object, of course, it's necessary to know not only its bearing but also its range. While direction is easily accomplished and requires only an eye, distance is a bit trickier and requires two eyes. Binocular vision gives us a sense of depth perception in much the same way that binaural hearing gives us directional hearing. The signal reaching one eye is slightly different from that reaching the other eye according to what's called parallax. Look at your extended thumb first through one eye, then the other. You'll notice that its position against its background seems to shift as you switch eyes. This is parallax, and it's due to the marginally different vantage points of your two eyes. The brain actually receives two channels of visual input that vary just a bit by virtue of parallax. It subconsciously compares these two images and calculates the distances of objects within them based on their different relative positions.

Three-dimensional movies create the illusion of depth by getting the brain to work the same way. They're really two movies that were shot at the same time through two cameras separated like two eyes and then shown at the same time through two projectors. One of these movies is exclusively for the right eye, the other for the left eye. In order to make sure each eye sees only the movie intended for it, it's necessary to equip each with a special lens that allows in only one of the two movies. While watching a 3-D movie the brain actually receives two marginally different movies at the same time, much as it normally receives two marginally different versions of reality.

II.21 | THE BEHAVIOR OF LIGHT

Although light and sound exist, travel, and are perceived in ways that are superficially analogous, they're two fundamentally distinct types of energy whose differences far exceed their similarities. In its essence, sound is the product of vibrating objects, the sloshing about of an elastic medium of transmission and the synchronized collisions of molecules. In every aspect it's dependent upon matter in motion. Light isn't. It depends upon matter only for its initiation. Matter, excited by either incandescence or fluorescence, is a necessary precondition for the existence of light. But light, once generated, can exist and travel independently of any material means. Light energy can and does exist even where no matter is present.

Light can also exist and travel in the presence of certain substances. Some solids, like diamond and glass, a few liquids, like water, and most gases are capable of conducting light. Such substances are said to be transparent. But light doesn't need these transparent media in order to get from one place to another. The light from the sun, for instance, reaches us only after traversing unimaginable stretches of empty space.

Light, in fact, operates best when matter is absent. Light traveling through transparent substances eventually loses its identity and relinquishes its energy as heat to the medium through which it's passing. Water, no matter how pure, glass, no matter how clean, and even air absorb small portions of the light passing through them. Light passes through these substances not because of their presence, but in spite of it. But a vacuum, being void of matter, is ultimately transparent. Light traveling in a vacuum, uninhibited by any sort of material interference, will never give up its energy or lose its identity. It will never deteriorate into anything else, but remain forever light.

Sound, dependent as it is upon the movement of matter for its propagation, can travel only as fast as the elasticity of its conducting medium will permit. Light, unhampered by any material encumbrances, is apparently capable of leaping from place to place instantly. Certainly there's no need to design cinemas or ballparks so as to prevent the blurring of light. There's no danger of light reverberation or light echo. And, for now anyway, there's no serious thought given to the possibility of flying at speeds beyond the light barrier. For most earthly purposes light travels at an infinite speed, is capable of getting from point A to point B in no time at all, and can be everywhere at once. Generally speaking, the behavior of light allows us to see things as they happen without any sort of time lag or distortion.

For thousands of years the infinite speed of light was considered literally true. It was so obvious that no one gave it a second thought. No one except Galileo, that is. Once again it was that defiant old Italian who first questioned the conventional wisdom. He boldly asserted that light was a physical entity of some sort and therefore had a finite, physically fixed speed. He even went so far as to attempt to measure this speed. Due to a lack of adequate equipment he failed miserably in this endeavor, but he did succeed in popularizing the idea that light has a speed.

We now know for certain that light has a speed, and we know what it is. It's somewhere around 300,000 kilometers (186,000 miles) per second in a vacuum. Through air it travels slightly slower—through water, glass, and other transparent substances slower yet. So light, like sound, has many different speeds depending upon what it's traveling through. But all these speeds, even though the slowest of them lies well beyond comprehension, are finite. Light doesn't, as the ancients believed, get from one place to another in no time at all. It needs time to cover distance.

There are two distinct aspects of the speed of light that are of interest and considerable consequence. The first—the fact that light has a very real, although incredibly swift, velocity—is the basis of most modern scientific thought, including the theory of relativity, a subject that will be discussed a bit later. (See Units V.7–10.) The second—the fact that light has different speeds in different environments—has more familiar ramifications, because light that changes speed may also change direction. This is called refraction.

Refraction occurs when light leaves one transparent substance and enters a second transparent substance in which it has a different speed. Or, in other words, when light passes between substances with different optical densities. A vacuum, air, window glass, water, and diamond all have different optical densities, and light refracts when it passes from one of these to another. What's more, air, water, and most other transparent fluids also have variable optical densities depending upon their temperature. Light passing from

■■

Light vs. Sound

The speed of light is so immense that it defies any attempt to illustrate it meaningfully. A flash of light can make over 30 round-trips between New York and Los Angeles in less than a second, and for covering any commonly encountered distance requires next to no time at all.

Because of this blinding speed a lightning bolt is seen virtually as it happens regardless of one's vantage point. Its coincident thunder, meanwhile, requires a noticeable amount of time to reach various ears depending upon their location. This disparity in travel times can be used to approximate the distance of a thunderbolt. Neglecting the travel time of the light, every second between lightning and thunder corresponds to 343 meters of distance covered by the sound. Three seconds equals 3×343 meters or just about one kilometer; five seconds about a mile.

To try and put this in some sort of perspective, imagine that it's the light that takes five seconds to reach an observer. In this case the thunder wouldn't be heard until over a month and a half later.

■■

cold water to hot water will refract just as surely as will light passing from glass into air.

Refraction can be a nuisance. Trying to fish a coin off the bottom of a swimming pool is difficult because the coin isn't really where it seems to be. Refraction bends the light connecting the coin to your eye as it passes from the water into the air and thus distorts the line of sight. Refraction can be beautiful. A diamond's sparkle is more dazzling than that of cut glass because it's a better refractor of light. Refraction can also be a source of deception. Desert mirages are the result of images being misplaced by the refraction that results when light passes from hotter into relatively cooler air. Many UFO sightings can be explained the same way, as can those imaginary puddles on the road on a sunny day. They're just parts of the sky put in the road by refraction. And refraction can be romantic as well. Light from the far-flung reaches of the universe is refracted as it enters the atmosphere and penetrates layers of air at various temperatures. This is what makes the stars twinkle.

Finally, refraction can be useful. If you're reading these words through a pair of eyeglasses or contact lenses you have refraction to thank. Lenses of all types reduce, magnify, focus, and disperse optical images through the scientific application of refraction. Microscopes, telescopes, binoculars, bifocals, and cameras all depend upon the bending of light as it passes from the air into glass and then back into air or other types of glass to accomplish their given tasks.

II.22 LIGHT WAVES

Light, in many important regards, bears little if any resemblance to sound. Sound travels by means of matter. Light travels as pure, untouched energy. Sound plods along at a leisurely pace. Light flashes across space in nothing flat. Sound depends upon the elasticity of its conducting medium. Light requires no medium at all. Sound spills around obstacles and fills available space as it remains captive in its medium. Light travels unerringly in straight, narrow paths and bends precisely when refracted.

Obviously, then, the physical format of light must be very different from that of sound. Light, in fact, is very different from any other known natural phenomenon, be it energy or matter, macroscopic or subatomic. The mechanism of light can't be properly explained in terms of any common objects or experiences. The nature of light is unique, complex, and abstract—enough so to have been the subject of one of the longest-running and liveliest debates in the history of science, one that engaged every great mind from Isaac Newton to Albert Einstein. (See Unit V.11.) Only quite recently has the nature of light been satisfactorily explained by the experts to the experts. And even this explanation isn't one that can be visualized, but rather one that can only be mathematically expressed.

Although the proper understanding of light requires knowledge, imagination, and scientific sophistication that few of us possess, much of its behavior can be adequately discussed by way of a single, sweeping simplification. This simplification holds that light, like sound, travels as waves. But think about it. If light can travel in the absence of any material conveyance, just what is it that undulates? The idea of a wave existing in the absence of any physical motion is pretty tough to harbor. It's sort of like a box with no shape or a speed with no direction. No, light waves aren't waves in any conventional sense. They, like many other aspects of the natural world, must be taken on faith alone.

Light, casually, crudely, and for the purposes of convenience, is usually pictured as existing in waves that resemble water waves. Transverse waves, the kind seen at the beach or in the bathtub, adequately explain much of the behavior of light. A transverse wave consists of a uniform series of equally spaced crests and troughs moving in the direction of the energy flow. The distance from the top of one crest to the bottom of the adjacent trough is the wave's height or amplitude. The distance from one crest or trough to the next is the wave's wavelength. One full wavelength consists of one complete trough and one complete crest. The passage of one full wavelength is called a cycle. The number of crests or troughs passing any given point in one second is the wave's frequency, expressed in cycles per second.

Because all light waves travel at the same speed, the wavelength and frequency of light waves are rigidly linked in a precise mathematical interrelationship. Waves of assorted wavelengths, like a group of joggers with different-

The Speed of Light

The first attempt to measure the speed of light was made by—who else?—Galileo, who flashed lantern signals back and forth between distant hilltops. Although he was unable to get results that were reliable, he did manage to inspire Olaus Roemer (1644–1710).

Roemer was a Danish astronomer fascinated by Io, one of the moons of Jupiter discovered by Galileo in 1610. Night after night he watched Io spellbound, keeping a detailed record of each time it disappeared behind one side of Jupiter and of when it emerged again from the other.

After months of collecting data he determined that the elapsed time of Io's orbits fluctuated by a total of about 22 minutes. This was a startling fact, for it ran against the accepted notion of the universe as a highly ordered, utterly regular mechanism. Somehow, Roemer had to explain the disturbing irregularity of Io's orbit about Jupiter.

He knew that the distance between the earth and Jupiter varied by about the size of the earth's orbit. He knew, too, that the elapsed time of Io's orbits routinely seemed to shrink as the earth neared Jupiter and stretch out again as the two planets drifted apart.

Roemer concluded that the fluctuation in orbit times resulted from the varying amount of time needed for the light from Jupiter to reach him on planet Earth. Using estimates of the distances and times involved he then calculated the speed at which light evidently traveled. It turned out to be fantastically huge. Due to a fortunate cancellation of errors it also ended up being fairly accurate.

Regardless of his results, though, Roemer was the first to provide physical evidence of the finite speed of light. The year was 1676.

—How do you suppose Galileo conducted his lantern experiment? What caused its failure?

—Now, with better clocks, we know that the actual observed fluctuation in Io's orbit is about 16 minutes. Was Roemer's estimate of the size of the earth's orbit too big or too small?

—Is there any way to see what is happening on Jupiter at this very instant?

—What would an observer near the North Star, 460 light-years away, be seeing on earth right now if he or she had a powerful-enough telescope?

■■■■■■■■■■■■■■■■■■■■■■■■■■■■■■■■■■■■■■■

Brightness

Brightness is usually measured in units called foot-candles, one foot-candle being based on the brightness of one candle at a distance of one foot. A 60-watt bulb equals about 60 candles, and reading at a distance of one foot from such a bulb would provide you with about 60 foot-candles of light. Direct sunlight provides up to 1,000 foot-candles of light and more. Considering its distance, this means that the sun burns with the brightness of 2.5 billion billion billion candles.

The ability to see depends utterly upon adequate light. Demanding visual tasks like sewing require as much as 200 foot-candles of light. Reading can be comfortably done in about 20 foot-candles of light, dining in as little as 3 foot-candles. Trying to see in insufficient light taxes the body and can lead to eyestrain, headaches, dizziness, and even indigestion.

In very dim light vision deteriorates rapidly. Detail discrimination, depth perception, and even color vision suffer greatly. As Ben Franklin put it, "In the dark all cats are gray."

■■■■■■■■■■■■■■■■■■■■■■■■■■■■■■■■■■■■■■■

sized strides, all cover the same amount of distance in the same amount of time. Waves with shorter wavelengths need to have higher frequencies the same way joggers with shorter strides need to have faster cadences. Light waves, then, can be of a low-frequency/long-wavelength nature, a high-frequency/short-wavelength nature, or of some other intermediate nature. The particular frequency/wavelength nature of any individual light wave is what determines its color. Its amplitude, meanwhile, determines its energy content or brightness.

Light waves are rarely if ever encountered on an individual basis, but almost always en masse. The light from the sun, a light bulb, or just about any other luminous source is a jumble of waves of all sorts of frequencies and intensities. The brightness of light in this form depends not so much upon the amplitudes of the individual waves involved, but more on their raw numbers. And color under these conditions is the result of how many waves of each wavelength are present. When all are represented to a more or less similar degree the result is white light.

In many regards all light waves behave similarly regardless of their wavelength or amplitude. All waves, for example, are conducted identically through a vacuum. White, random light, like sunlight, thus remains a white mixture as it travels under these conditions. The waves of all frequencies are conducted at the same speed and all retain their identities equally for the duration of the journey. Waves of all frequencies are also conducted at the same speed in a transparent medium. During such a voyage they all deteriorate into heat more or less equally.

But waves of different frequencies don't always behave identically. If they did they would always remain mixed and nearly everything would appear white, black, or one of many shades of gray. Because they don't the world appears in a kaleidoscope of color.

II.23 | THE PERCEPTION OF COLOR

Serious thinkers and idle speculators alike have wondered what makes a cherry red, grass green, and the sky blue ever since the beginning of recorded time and probably long before that. The first organized attempts at an explanation we know of were made, once again, by the Greeks. Empedocles thought that everything emitted steady streams of tiny particles that struck the eye and thereby produced the perception of color and the sensation of vision in general. A cherry gave off red particles, grass gave off green particles, and the sky gave off blue particles continuously and endlessly. Just where this inexhaustible supply of particles came from he didn't say. Aristotle, who, you may recall, was not too enamored of tiny particles, attributed the existence of colors to the air. Again, the particulars of the scheme were left somewhat undeveloped.

Perhaps because of the obvious shortcomings of these earliest efforts, other theories abounded. Most of them were pretty farfetched, a few downright ridiculous. They all agreed, though, that an object's color, like its shape, size, and weight, was an innate part of its identity that existed independently of its being seen. Color was a function of the observed, not the observer.

The mechanism of color remained the subject of repeated and wild conjecture until 1665. In that year Newton set his considerable mind to the task and performed a number of now-famous experiments. The experiments themselves were almost childishly simple. The conclusions he drew from them changed forever the way in which light and color were regarded.

Newton merely passed a beam of common white light through a glass prism and noted, as had many before him, that the light that emerged from the prism wasn't white, but a rainbow-colored band. From this simple fact he deduced, as none before him had, that white light wasn't truly white, but only appeared that way. White light, he said, wasn't a homogeneous collection of rays each of which was white, but rather a heterogeneous assortment of rays of many different colors. None of white light's component rays, in fact, were truly white. No white light emerged from the prism. It was only the total effect produced by the mixture of many separate colors that gave the impression of whiteness. This he proved by passing the rainbow of light produced by one prism back through a second prism from which it emerged again combined as a single beam and again visibly white. Newton, in effect, disassembled a beam of white light into its component colors, then reassembled these colors to reconstruct the beam of white light.

White light, Newton demonstrated, doesn't really exist in a strictly physical sense. Every individual ray of light has a color, but none is actually white in and of itself. White light exists only in a physiological sense—only in the eye of the beholder. White light is a subjective experience and not a natural reality. Our eyes see light as being white only because they're unable to dissect light into its component parts the way a prism does. The reds, greens, and blues present simultaneously in a beam of sunlight overlap in our visual experience to produce the sensation of white light. Color, contrary to what had always been assumed, isn't a function of what's being seen so much as it's a function of what's doing the seeing.

The colors in which light actually exists are those into which a prism fractures a beam of sunlight. This multicolored spread of colors is properly called the spectrum. Because sunlight contains every possible wavelength of visible light the spectrum contains every possible color of light. Every wave has its own particular wavelength, and that wavelength determines its color—precisely and unalterably. All light waves of the same wavelength are of the same color, and all light of the same color has the same wavelength.

A prism takes a beam of white light and sorts it into its component waves according to their wavelengths. The longest wavelength of light in this hierarchy corresponds to a deep red, the shortest to a vivid purple. In between these extremes lie various hues of orange, yellow, green, blue, and all the countless intermediate shades. The prism produces this result because not all wavelengths of light behave the same when being refracted—when passing from one substance to another of different optical density. Each separate wave bends to a degree determined by its wavelength or frequency. High-frequency violet light bends the most; low-frequency red light bends the least. White light entering the prism is all traveling along the same path. Refracted light leaving the prism, having been bent to various angles, is fanned out in a spectrum of light graded according to wavelength and color.

Without doubt the most spectacular display of this sort is a rainbow. A rainbow results when white sunlight is refracted by millions of suspended water droplets. Each droplet acts both as a tiny prism and as a tiny mirror. Light entering each droplet is refracted as it passes from air into water. This refracted light is then bounced off the back of the droplet and back into the air, being refracted a second time as it does so. The double dose of refraction is enough to separate the light according to wavelength. Each droplet throws only one small portion of its spectrum on any observation point. Each droplet is thus seen as being of only one certain color depending upon the vantage point from which it's being viewed. The angle formed by the sun, the droplet, and the observer determines what color a particular drop will appear. But within the sky are many droplets. Each of these forms its own angle and each is seen in its own color. Taken together they form the vast, sweeping, multicolored arc that, no matter how many times it's seen, is always something special.

II.24 REFLECTION

All light waves, no matter what their frequency, cruise along identically through a vacuum or anything transparent. When they pass from one transparent medium to another of a different optical density, though, light waves behave in a manner that depends upon their frequencies. Each is bent to a degree determined by its frequency or wavelength. This nonconformity of light waves as they're refracted is what gives us rainbows. Light waves also behave differently by wavelength when subjected to reflection. This is what makes cherries red and grass green.

Cherries and grass, unlike raindrops, transmit no light. They're opaque, not transparent. Light that falls on an opaque object, unable to penetrate it, can do one of two things: it can be absorbed by the object or it can be reflected by the object. Whichever of these two actions any particular light wave takes is determined by its wavelength and the nature of the opaque object upon which it lands. Every opaque object reflects its own particular assortment of light waves and absorbs the rest. The light waves that an opaque object absorbs warm it slightly. Those that it reflects illuminate it and make it visible.

At one extreme of the opaque scale are those objects that absorb almost no light and reflect a lot of it. In extreme cases, the quality and quantity of light bouncing off these objects is just about the same as that which falls on them. If these objects are highly polished, in fact, the light coming off them is nearly indistinguishable from that illuminating them. Mirrors, mountain lakes, and chrome-plated coffeepots redirect patterns of light while leaving them pretty much intact. When you see yourself in a mirror you're viewing a set of light waves that illuminated your face, traveled in formation to the mirror's surface, and were then reflected nearly undisturbed back into your eye.

Other highly reflective objects don't have such smooth surfaces. They reflect most of the quality and quantity of the light falling on them, but scramble it in the process. These objects, because they reflect all the light falling on them, take on the color and the intensity of the light by which they're illuminated. When illuminated by white sunlight they appear white. If illuminated by blue light they appear blue. But, because their surfaces are uneven, they rearrange light waves falling on them and are unsuitable as mirrors. Movie screens, typing paper, and almost anything else that normally appears white but rather dull fit into this category of opaque objects.

Unlike mirrors and movie screens, most opaque objects reflect light rather selectively. These objects take on the color of the light they reflect. Cherries absorb all wavelengths of light except red, and appear red even when under white sunlight. Grass similarly reflects only green light. A school bus reflects yellow light. Usually an object reflects more than just one type of light. The assorted wavelengths of light reflected by these objects fuse in the eye to yield all sorts of complex hues that, like white, don't appear in the spectrum. The number of these subjectively experienced colors—mauve, chartreuse, lavender, etc.—is theoretically limitless.

There are also objects that reflect practically no light at all. These objects exhibit an absence of color and appear black. Any object, in fact, appears black when not illuminated by a color of light it's capable of reflecting. Cherries, when lit by blue, green, or any other color of light lacking red wavelengths, appear as black as lumps of coal.

Any object big enough to be seen assumes the color of the light it reflects. Objects too small to be seen, of course, are colorless as well as invisible. Light waves simply wash over microscopic particles the way an ocean wave washes over a small stone. The wave proceeds undisturbed and the presence of the stone goes unnoticed. Objects of this dimension remain invisible even under the most powerful microscopes because there's simply no reflected light to magnify. A zero-sized image, blown up a million times, is still a zero-sized image.

At the threshold of visibility are objects that are just barely able to reflect light. These objects don't reflect enough light to take on any detail or meaningful image, but just enough to be detectable. Light waves that strike these marginally visible bits and pieces don't bounce off cleanly, as they do off of larger objects, but are merely disrupted as they pass by. Objects of this magnitude, like specks of dust or tiny water droplets, are properly said to scatter rather than reflect light.

According to the Tyndall effect, named after John Tyndall (1820–1893), the British scientist who first described it in 1869, objects at the threshold of visibility scatter light selectively depending upon their size. The smallest of these particles scatters light of the shortest wavelength, and progressively larger objects scatter progressively longer waves of light. A collection of like-sized particles scatters light of only a particular wavelength, or color. A collection of particles of various sizes scatters light of various colors. The assorted sizes of water particles in a cloud scatter assorted wavelengths of light that combine in the eye to produce the color white. Black clouds result when the water particles get so large that they block out rather than scatter overhead sunlight.

It's the scattering of light by impurities in the air that makes it possible to read a book in the shade of a beach umbrella. Without the Tyndall effect the direct sunlight surrounding the umbrella would be brilliantly white and the shadow beneath it totally black. Twilight and dawn depend upon scattered light as well. Without it we would have full daylight right up until sunset and then be suddenly thrust into darkest night, just as if somebody had flipped a huge light switch. During the day it's the Tyndall effect that illuminates the sky and obliterates the stars. Without it the daytime sky would look just like the nighttime one with the addition of one huge extra star—the sun. The sky itself would remain unilluminated and black.

Unlike the water droplets in a cloud, the particles suspended in clean air aren't a random assortment of sizes. Air particles are predominantly rather smallish. Most of the light they scatter is therefore of a very short wavelength. Specifically, the air most effectively scatters blue light. It's blue light that ricochets around the sky and finally lands in our eyes. We experience the sky according to this scattered light. We see it as blue.

While the blue portion of sunlight illuminates the sky, the rest of it passes on through the atmosphere to illuminate the surface of the planet. The sunlight we live by, having had its blue elements leached out by the sky, takes on a yellowish hue. The sun itself appears not white, as it would from a vantage point above the atmosphere, but yellow. At sunrise and sunset this effect is exaggerated because the sunlight has to penetrate the atmosphere at a shallower angle and thus pass through more of it. At these times even more of the shorter wavelengths are subtracted from the sun's full spectrum of light. What's left appears orange or even red.

II.25 ELECTROMAGNETIC RADIATION

Visible light is composed of light waves of many different wavelengths, all of which lie within very strict parameters. At one boundary is the longest visible wavelength of red, at the other is the shortest visible wavelength of violet. These wavelengths, however, encompass only a small fraction of all possible wavelengths. Waves also exist that are longer than those of the reddest red and shorter than those of the most violet violet. These waves are identical to light waves in every aspect except one: they can't be seen. But to think of these outlying waves as invisible types of light is incorrect. They aren't light that the eye can't see. Light, rather, is a special sort of wave that the eye *can* see. All such waves, whether humanly detectable as light or not, are properly known as electromagnetic radiation. (See Figure II.4.)

All electromagnetic radiation travels in the same wavelike fashion. All these waves move at the speed of light and can be transmitted through a vacuum. They can also be transmitted through various transparent media, although just what constitutes a transparent medium depends upon the wavelength under discussion. Some are effectively blocked by a piece of tin foil while others can penetrate a slab of lead. The many forms of electromagnetic radiation are also subject to refraction, reflection, and absorption in much the same way as are waves of visible light. But the total spectrum of electromagnetic radiation extends far beyond the limits of visible light, and so do the behaviors of the waves making up this larger spectrum.

Just beyond the longest waves of visible red lies infrared radiation. Infrared radiation is the only nonvisible form of electromagnetic wave that can be sensed by the human body. It's not seen, but felt as heat. The heat coming off an open flame is infrared radiation. Almost half the energy radiated by the

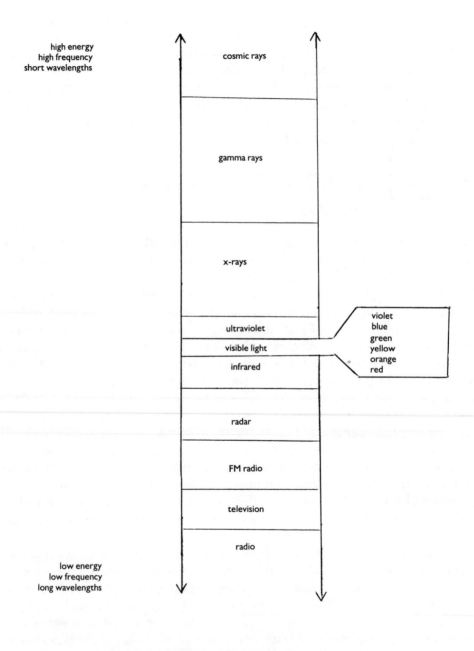

Figure II.4   The electromagnetic wave spectrum

sun and up to 95 percent of the energy radiated by an incandescent light bulb is also infrared radiation. Whenever heat is transported via radiation, in fact, it's done so in the format of infrared radiation. (See Unit II.15.) Whenever heat appears in the company of light you can be pretty sure this extralong red wave is present. But heat can also be radiated in the absence of light. The heat coming off a radiator, as well as the heat given off by a warm-blooded animal, travels on waves of infrared electromagnetic radiation.

Although invisible to the human eye, infrared radiation can be detected by special optical devices and even recorded on special types of photographic film. Using these techniques infrared radiation can actually be "seen" by indirect methods. What's more, it can be seen in the total absence of any visible light. A sniper with an infrared scope on his rifle can pick out a body-temperature target from its cooler surroundings even on the darkest night because of the infrared radiation it emits. A spy plane or satellite with infrared cameras on board can record the heat given off by an underground nuclear bomb test. The thermal pollution of power-generating plants is tracked the same way. And a thermograph can detect and locate the heat given off by a tumor growing deep inside the human body.

Waves longer than those used to carry heat can't be seen or felt, but can still be detected with the assistance of special equipment. The most common use of these waves is in the operation of radar systems. Electronically generated radar waves scan the sky in much the same way as does a searchlight. When they encounter a reflective object, like a plane, they bounce back from that object to their source where they're seen by electronic eyes. The direction of the reflected beam gives the plane's bearing. The time lapse between the transmission and reception of a radar wave is used to calculate its range. Radar has several advantages over visible light for this purpose. First, it can travel far greater distances and, second, it can penetrate clouds, fog, and rain.

Beyond radar waves lie other types of waves with even longer wavelengths and lower frequencies. These waves, too, are capable of traveling long distances regardless of weather conditions or time of day. Waves in this category, called microwaves, have few known natural sources or functions, but can be artificially generated and put to good use. They're most commonly employed to transport radio and television signals. There also exist waves with lower frequencies yet, but these have no practical applications nor any known natural functions. There's no theoretical limit as to how long a wave can be, but eventually it loses its identity as a wave and its ability to transport energy.

Besides all the waves too long to see there are those that are too short to make any impression on the human nervous system. Just beyond the fringe of visible violet light lies ultraviolet radiation. The sun radiates vast amounts of this stuff. Most of it's thwarted by the atmosphere, where it's either reflected back into space or absorbed and turned into heat. That which penetrates the atmosphere is powerful enough to cause sunburn. Because ultraviolet radiation can penetrate clouds, it's quite possible to get burned by it even on an overcast

● ●

Microwaves

The microwave portion of the electromagnetic spectrum has been artificially divided into many thin slices. Each of these slices contains waves of very specifically defined frequencies. There are slices for AM (amplitude modulation) radio, FM (frequency modulation) radio, short-wave radio, citizen's band (CB) radio, and cellular and long-distance telephone transmission, among numerous others. Each of these is further divided into razor-thin slices corresponding to specific channels. Various intercoms, garage-door openers, and remote-control devices further clutter these frequencies.

A less organized blast of these very same waves heats up yesterday's leftovers. Intensity is the only difference. Whereas the 50,000 watts of microwaves being broadcast by a television station are dispersed over huge expanses of sky, the 500 watts or so of microwaves generated by the typical oven are cooped up inside a small steel box.

The microwaves are generated by a delicate device called a magnetron. This ray-gun sprays microwaves into the oven where they bounce off the reflective walls and door, eventually converging in the center of the oven where they strike something, a pork chop for instance, that absorbs them. As does all absorbed radiation, the microwaves then turn into heat, in this case enough of it to cook the meat. Were TV and radio signals powerful enough, they too could cook. As it is, they're too weak to even be detected except by sophisticated electronic equipment.

Whereas a conventional oven cooks food from the outside in, a microwave oven cooks it everywhere at once as the radiation penetrates it. This accounts for the speed of microwave cooking. Besides being fast, microwave ovens are also efficient because they heat up only the food, not the whole appliance and the rest of the kitchen. Microwave ovens are also popular because they're so clean.

—Is there a limit to the number of television and radio channels that can be broadcast?

—Why do some microwave ovens have turntables that rotate the food as it cooks?

—Why is it dangerous to run a microwave oven when it's empty? Why should you avoid putting reflective items, like knives and forks, in a microwave oven?

—Why does it take so much longer to heat up four slices of pizza in a microwave oven than a single slice?

● ●

day. Artificially generated ultraviolet radiation is used in sunlamps to produce safe tanning. It's also employed to sterilize medical instruments and has numerous manufacturing applications, usually as the energy source for certain chemical reactions. It's natural ultraviolet radiation that triggers the chemical reactions that transform exhaust and other common pollutants into the noxious stew of more complex chemicals known as smog.

Beyond ultraviolet radiation are waves with even shorter wavelengths and higher frequencies. The first class of these waves was discovered quite by accident in 1895 by a German scientist, Wilhelm Roentgen (1845–1923). He didn't have any idea what they were, and so he called them x-rays. We now know that they, like light and radio signals, are a type of electromagnetic radiation, but their mystery-laden appellation has endured despite efforts to rename them Roentgen waves.

The most common employment of x-rays, of course, is in medicine. Within weeks of their discovery they were being used to photograph broken bones. An x-ray photograph results when a blast of x-ray radiation is beamed through a body and onto a special film behind it. Some parts of the body are transparent to the waves and the film behind them is exposed to the radiation. Other parts, like the bones, absorb the radiation and so throw a shadow on the film.

X-rays have other valuable applications as well. In industry they ensure quality control by revealing flaws in manufactured items the same way they reveal breaks in a fibula or a tibia. Roentgen radiation is used in science and industry to examine objects too small to be seen by more conventional means. Some particles too small to reflect waves of visible light are big enough to reflect the shorter waves of Roentgen radiation. Even objects visible in normal light can be seen in more detail when examined with shorter waves.

Waves far shorter than x-rays also exist. They're of two general types: gamma rays and the even shorter cosmic rays. Gamma rays are the product of radioactive decay and nuclear reactions. (See Unit II.29.) Cosmic rays originate from as yet not totally understood sources in outer space. Both have enormous penetrating power and are potentially quite dangerous. Fortunately, the atmosphere shields us from cosmic radiation and the amount of naturally occurring gamma radiation is, for the most part, inconsequential. These rays, because they're so hazardous, have few uses outside of scientific investigation.

II.26 | THE MAGIC OF MAGNETISM

As their name indicates, electromagnetic waves are the result of an interplay between electric and magnetic forces. Both magnetism and electricity operate by means of what are called force fields. It's these invisible, empty, intangible fields of force that allow electromagnetic radiation to exert its influence across the expanses of space. Just as a ray of light can traverse empty space, so a magnet attracts bits of metal with which it has no direct contact and an electric spark leaps from place to place without the benefit of any material pathway. In each case a force exists and acts as pure energy without the involvement of matter.

Electromagnetic waves are neither magnetism nor electricity, but a complex hybrid of the two, pulsating first as an electric force, then as a magnetic one. The exact nature of this hybridization is obscure. What is interesting, though, is the fact that the two seemingly unrelated phenomena of electricity and magnetism are really two aspects of a single, larger phenomenon.

The magic of magnetism and the enchantment of electricity were both discovered and explored individually long before any connection between them was realized or even suspected. The first knowledge of magnetism was recorded by the Greeks. In an area of Asia Minor they called Magnesia they discovered some unusual rocks that clung to metal objects and to one another. The rocks were pieces of iron ore we now call magnetite, and were natural magnets. Knowledge of electricity dates back to the same era. Around 600 B.C. Thales of Miletus rubbed pieces of amber with a cloth and then noted that the amber had the power to attract feathers in much the same way that rocks from Magnesia attracted bits of metal. The amber, which the Greeks knew as *elektron,* attracted the feathers with static electricity.

Both magnetism and electricity remained natural curiosities with little value beyond amusement for hundreds of years. No one had any success in explaining them or using them productively. And no one suspected their importance. Then, probably a thousand years or so after it was discovered, magnetite became much more than a toy. An oblong piece of the ore, when suspended by a string, was found to always assume a north-south orientation. And always the same end of the bar pointed to the north. Magnetite became widely known as leading stone, or lodestone, and was put to use for navigational purposes. With one end of a suspended leading stone pointing ever northward, all other directions could also be told. These first primitive compasses helped early mariners retain their bearings even on a cloudy night, and helped make possible the daring feats of Christopher Columbus, Ferdinand Magellan, and Sir Francis Drake.

But even after assuming great importance and widespread popularity, lodestone's operation remained a total mystery. It wasn't until 1600 that Dr. William Gilbert (1540–1603), the brilliant personal physician of Queen Elizabeth I, conducted a series of systematic experiments that formulated the principles of magnetism. Gilbert demonstrated that every piece of lodestone, or every magnet, had two poles: a north and a south. The north pole pointed always north when given the opportunity, the south pole south. According to the law of polarity, which Gilbert developed, the like poles of any two magnets repel one another while their opposite poles exhibit mutual attraction.

Gilbert was also the first to realize that the earth itself acted as a huge magnet and that it was the interaction of the global magnet with suspended pieces of lodestone that accounted for the north-south orientation of compass needles. In

■■■■■■■■■■■■■■■■■■■■■■■■■■■■■■■■■■■■

Two North Poles

The magnetic pole toward which the north end of a compass needle points is, in accordance with the laws of polarity first expressed by Gilbert, actually a south magnetic pole. There is, then, a south magnetic pole near the north geographic pole and a north magnetic pole near the south geographic pole.

The coincidence of the magnetic and geographic poles, while fortunate, is only approximate. The south magnetic pole, for instance, currently lies somewhere in northern Canada, hundreds of kilometers from its geographic counterpart. This is destined to change, however, for, as the use of the word *currently* would indicate, the magnetic poles have a tendency to move around a bit. Although continuous, their wanderings are slow and never take them too far away from the world's top and bottom.

Nevertheless, to find true geographic directions with a compass it's necessary to use a correction chart that indicates the difference in bearing for true north and magnetic north for any particular location. Because of the drifting of the magnetic poles these charts need to be periodically updated.

■■■■■■■■■■■■■■■■■■■■■■■■■■■■■■■■■■■■

accordance with Gilbert's law of polarity, the earth's magnetic north pole attracts the south poles of all other magnets and its south pole attracts the north poles of all other magnets. The magnetic north pole of a freely suspended compass needle always seeks the magnetic south pole of the earth, and its south pole strains ever toward the earth's magnetic north pole. Fortunately, these global magnetic poles coincide more or less with their geographic counterparts—opposites really—to provide a reasonably accurate approximation of direction.

The magnetic field generated by the global magnet has another less useful but more spectacular effect. The sun, in addition to throwing off visible, infrared, and ultraviolet radiation, also emits tiny, charged, subatomic particles—protons and electrons to be precise. Some of these particles stream toward planet Earth and enter its magnetic field. Once in the clutches of this force field they're funneled toward the magnetic poles. As they enter the atmosphere they interact with molecules of air in much the same way an electric current interacts with the neon gas trapped inside a "Bud on Tap" sign. The result is a red and green glow visible at night near the poles, especially during times of solar flares when the particles are more numerous. This display of natural fireworks is called the aurora borealis or northern lights in the Northern Hemisphere and the aurora australis in the Southern Hemisphere.

The exact mechanism by which magnetism operates has yet to be definitively explained. Most theorists believe it has something to do with the alignment of atoms within a piece of metal or magnetic material. Each atom, it's supposed, is a tiny magnet all its own. In nonmagnetic objects each atomic magnet points randomly in every conceivable direction and their force fields nullify one another. In a magnet, though, all the atomic magnets are aimed the same way. Their individual force fields reinforce one another and produce a magnetic field of obvious magnitude. Just what makes an individual atom behave like a magnet is a further mystery, but the best guesses attribute the property to the arrangement and motion of its electrons.

II.27 STATIC ELECTRICITY

Not surprisingly, magnetism's eerie cousin—electricity—also has as its source that tiny outrider of the atom, the electron. Under normal, electrically neutral conditions electrons are distributed throughout matter in a density that's approximately matched by that of protons. Each negatively charged electron is, in effect, paired off with a positively charged proton. The equal but opposite charges of each electron-proton couplet cancel one another and the net electrical effect is zero. There is, for all practical purposes, no electrical activity or potential.

But the density of electrons isn't always matched by that of protons. Electrons reside at a great atomic distance from their associated nuclei and are easily dislodged. Electrons stripped away from their nuclei become independent, negatively charged particles. The atoms they leave behind, meanwhile, are left with electron deficiencies and assume positive charges. Once dissociated, the negative electrons and positive atoms can move independently of one another. The electrons can congregate in one place and the atoms in another. Wherever the density of electrons exceeds that of atoms a negative charge is generated. Wherever the density of positive atoms is greater than that of electrons a positive charge is generated. Once such charges have been established there exists the potential for an electrical event.

The interaction of positive and negative charges drives electrical events in much the same way that the interaction of north and south magnetic poles drives magnetic events. Similar charges, like similar poles, repel one another. Opposite charges attract one another. And either a positive or negative charge attracts electrically neutral objects in much the same manner that either a north or south magnetic pole attracts neutral thumbtacks. All these behaviors are driven by the natural urge of electrons to establish and maintain more or less uniform distributions. They resist being crowded into negative charges or dispersed as positive ones, and do whatever they can to prevent such situations or alleviate them when they arise.

The ability of electrons to neutralize electrically charged circumstances depends upon the presence or absence of matter and the type of matter involved. Certain materials, called conductors, allow electrons to migrate freely and thus equalize any unbalanced densities. Other materials, called insulators, inhibit such action. The absence of matter acts as an insulator as well, but the insulating properties of a vacuum aren't as good as those of material insulators.

Most metals are excellent conductors. Electrons will readily flow along a thin metallic pathway like a wire for enormous distances. Salty or dirty water also conducts electricity fairly well. Standing in water during an electrical storm is dangerous because any lightning striking the water will be conducted to your body. In this case it's actually the dissolved impurities that provide the pathway for the electrons, for pure water is a good insulator. That's why you must be careful to add only very clean water to a car battery. The pure water allows the necessary chemical reactions to take place but prevents any internal discharge of the battery. Other good insulators include air, wood, glass, rubber, and plastics.

In the presence of conducting materials, electrical charges assume motion as hordes of electrons flow from dense, negative concentrations toward sparse, positive areas. This migration of electrons is what constitutes an electrical current, and it's while in this active, kinetic state that electricity is an obvious, powerful, destructive, or useful form of energy. In the presence of insulating materials, electrical charges remain isolated and motionless. While poised in these states of unbalanced charges electricity does very little. Kinetic electricity is usually called dynamic electricity and potential electricity is usually called static electricity.

Static electricity is what so fascinated Thales back in ancient Miletus. By rubbing amber with a cloth he transferred electrons from the cloth to the amber, thereby giving the formerly neutral cloth a positive charge and the formerly neutral amber a negative charge. The insulating air between the cloth and the amber prevented any immediate neutralization of these static charges. The negatively charged amber then attracted neutral, nonconducting materials like feathers. The same thing sometimes happens when you comb your hair. The comb picks up relatively loose electrons from your hair just as the amber picks up relatively loose electrons from the cloth. The negatively charged comb and the positively charged hair are attracted to one another. The hair rises toward the comb. The air between them prevents any neutralization from taking place. The effect is especially noticeable on dry days because dry air is a better insulator than humid air.

Eventually, static electrical forces discharge and neutralize one another, either because the insulation between them is reduced to a point where it can no longer hold them apart or because they grow large enough to overcome any intervening insulation. When you shuffle across a carpeted room you charge yourself with static electricity. Your shoes pick up millions of loose electrons from the floor. These electrons remain attached to you as long as you remain insulated on all sides by air. The discharge doesn't occur until you touch a doorknob, a light switch, or a water faucet. Any of these conductive objects breaks the insulating shield of air and allows the electrons to seek more neutral environments. They immediately do so, often with a sensational, visible, audible spark.

The discharge of static electricity can assume truly awesome proportions. Lightning, perhaps the most spectacular of all natural events, is the conse-

■■■■■■■■■■■■■■■■■■■■■■■■■■■■■■■■■■■■

Lightning

Benjamin Franklin (1706–1790) first demonstrated that lightning was nothing more than electricity with his famous kite-flying exploits in 1752. Later that same year he developed the lightning rod, which not only disarms lightning by guiding it harmlessly to the ground, but also actually helps to prevent it by providing a pathway for the gradual, nonviolent migration of electrons.

Despite the widespread use of lightning rods, however, lightning remains one of nature's most terrifying spectacles and is responsible for numerous deaths and untold destruction every year. On the positive side, lightning creates tons of nitrates from the oxygen and nitrogen in the air. These nitrates fall along with rain and are excellent fertilizers.

All lightning, by the way, is basically the same. Sheet lightning is just lightning that takes place so far within a cloud that its path is obscured. Heat lightning is nothing but common lightning that takes place so far away that its blinding bolt can't be seen nor its attendant thunder heard.

■■■■■■■■■■■■■■■■■■■■■■■■■■■■■■■■■■■■

quence of static electricity. As water droplets mill about within a cloud they rub against the air in much the same way as a swatch of cloth rubs against a piece of amber or a bedroom slipper rubs against a shag carpet. The electrons become unevenly distributed and are kept that way by the buffer of air. Eventually the attraction between positive and negative regions of a cloud, or between one cloud and another, or between a cloud and the ground, becomes great enough to overpower this insulting gap. The electrons explode through the sky as a bolt of lightning and the imbalance of charge densities is instantly rectified.

II.28 | ELECTRIC CURRENT

Static charges, whether generated deliberately with amber and cloth or naturally in a storm cloud, remained the only type of electricity known to early investigators. They could observe electricity in its motionless potential state or they could observe it during its blinding discharge. In the first case not much happened and in the second case whatever happened happened too quickly to make any sense out of. Understandably, not much progress was made toward deciphering the nature of the electrical beast or toward harnessing its fearsome power. Not until 1800, the year that an inventive Italian physics professor, Alessandro Volta (1745–1827), put a silver spoon and a piece of tin foil on his tongue, connected them with a copper wire, and noted a distinctly sour taste in his mouth.

What Volta tasted were the acids produced by the chemical reactions involving the silver, the tin, and his saliva. These were the first known examples of electrochemical reactions, the type of chemical reactions that take place in any battery. Volta's oral contrivance was, in fact, the very first battery. He proceeded to produce more conventional batteries by stacking alternating zinc and silver discs interspliced with paper soaked in saltwater. Known at first as voltaic piles, these contraptions produced a continuous stream of electrons rather than an instantaneous spark. They produced predictable and controllable currents of electricity rather than random flashes. Wild electricity had been captured and was soon to be tamed.

Using Volta's battery as their primary tool, other scientists, most notably the Frenchman André-Marie Ampère (1775–1836) and the German Georg Simon Ohm (1789–1854), quickly solved many of electricity's mysteries and described its behavior with mathematical formulas. Volta's battery was found to operate by means of two simultaneous chemical reactions. One reaction required the disposal of surplus electrons while the other required a supply of electrons. By connecting the two reactions with a conducting wire the electrons shed by the first reaction were fed directly into the second reaction, thus enabling both to proceed. The flow of electrons through the wire from one reaction to the other was what constituted the electric current.

The flow of electrons through a wire possesses characteristics analogous in several ways to the flow of water through a hose. The flow in each case has a certain pressure. In the case of water the pressure depends upon the power of a pump. In the case of electricity it depends upon the force with which electrons are spewed into one end of the wire and the hunger with which they're sucked out of the other. The driving force behind a flow of electricity is called voltage in honor of the fearless Italian who used to gauge its intensity by the brightness of the flashes it caused on his closed eyelids. Voltage depends upon the nature of the two electrochemical events occurring at the opposing ends of the wire, and the nature of these reactions in turn depends upon the types of materials involved. Volta's original silver, tin, and saliva have been replaced by nickel, cadmium, and strong acids in the quest for more vigorous electric currents.

The voltage driving any flow of electrons is offset to some degree by the wire through which the current travels. The resistance of the wire depends upon its diameter, its length, and the material it's made out of. Short thick wires, like short thick hoses, have low resistances. Long thin wires have high resistances. Wires made out of good conductors have less resistance than those made out of poor conductors. In any event, resistance is measured in units called ohms.

The relationship between the driving force of voltage and the hinderance of resistance results in a rate of flow. The rate of flow is called amperage and is measured in amperes, or amps for short. One volt of pressure will drive electricity through a wire with 1 ohm of resistance at the rate of 1 ampere per second. Electricity flowing at this rate is said to deliver 1 watt of power. The rate of flow can be boosted by either raising voltage or lowering resistance. A

● ●

Plugged In

These days, with everything from toothbrushes to can openers being driven electrically, the conversion of electrical energy into mechanical energy has taken on special importance. It can all be traced back to a lucky accident.

It happened on an April evening in 1820. Hans Christian Oersted (1777–1851), a Danish physics professor, was giving an otherwise routine lecture at the University of Copenhagen. While conducting an experiment for his class he inadvertently passed a compass near a wire that was carrying an electrical current. Oersted noticed that the compass needle was deflected from its normal north-south orientation and thus made what is frequently called the greatest discovery ever made in a classroom.

Oersted had stumbled upon what we now call the motor effect, the fact that electricity in motion produces a magnetic field and that any magnet—like a compass needle—brought into that field will be set into motion. That motion can and now frequently is harnessed and put to work. Most of our modern gizmos are powered by electric motors that are, in simplest terms, just magnets being continually deflected by an electrical current.

Ten years later, Michael Faraday (1791–1867), a British physicist, made a much more intentional but no less important discovery by, in effect, running Oersted's experiment backward. He mechanically rotated a magnet and thereby set up an electric current in a nearby wire. This conversion of mechanical to electrical energy is now called electromagnetic induction. This phenomenon is the basis upon which electrical generators operate—the generators that supply us with the power we use to brush our teeth and open our cans.

Together the motor effect and electromagnetic induction underline the relationship between magnetism and electricity. The oersted and the faraday, in fact, are now terms used to describe certain units of magnetism and electricity, respectively.

—How many electrical devices can you identify in your household? How many of them employ an electric motor? Which of these operate on AC and which on DC?

—What is the source of the electricity used in your home?

—How does a hydroelectric dam generate electricity? What about generating plants that rely upon coal or nuclear power?

—Electric cars are being promoted as a pollution-free means of transportation. What is the true impact of an electric car upon the environment?

● ●

force of 1,000 volts pushing a current through 1 ohm, for instance, results in 1,000 amps per second, or 1,000 watts, better known as a kilowatt. This rate of flow, sustained for an hour, delivers 1 kilowatt-hour of electrical power.

The current generated by a battery flows in only one direction. The electrons are always generated at one pole, called the cathode, by one chemical reaction, and always taken in again at the other pole, called the anode, by a different chemical reaction. This one-way flow of electrons is called direct current, or DC. A battery produces direct current whenever its two poles are connected by a wire or other conductive material. The flow of electrons from cathode to anode continues until the chemicals in the battery are totally consumed and the chemical reactions cease. Some batteries, like automobile batteries, are rechargeable. The recharging process consists of driving a current backward through the battery and reversing the chemical reactions involved, thereby replenishing the original chemicals.

Direct current, DC, is found wherever electricity is supplied by batteries: in flashlights, portable alarm clocks, remote-control race cars, and virtually any other electrical device that lacks a cord. Appliances that plug into a wall socket, on the other hand, operate on alternating current, or AC. Alternating current consists of an electrical charge flowing first in one direction and then in the opposite direction. Seemingly needlessly complicated, AC current is used for economic reasons. It's easier to generate than DC current, and it can be transported over long distances with less effort. The typical AC current reverses direction 120 times per second. Every two reversals result in one complete cycle of the current, and this type of electricity is therefore called 60-cycle power.

All electricity, whether AC or DC, is clean, odorless, relatively efficient, and easily converted to other forms of energy. Crock pots turn electricity into heat, microwave ovens turn it into electromagnetic radiation, light bulbs turn it into light, stereos turn it into sound, and electric motors turn it into mechanical energy. Electricity is usually produced from mechanical energy by means of generators that in turn can be powered by a variety of energy sources.

II.29 | NUCLEAR ENERGY

The work of Volta, Ohm, Ampère, Oersted, Faraday, and others provided a solid, practical base of knowledge about electricity. Its use grew steadily throughout the nineteenth century and its applications multiplied thanks to the genius of inventors like Thomas Edison. But all the while it was being produced in ever-greater quantities, distributed to an ever-increasing number of customers, and being put to ever more ingenious uses, no one really understood what it was.

Not until 1896 did Joseph J. Thomson discover the electron and thereby provide electricity with a satisfactory theoretical base. (See Unit I.26.) By that time electricity was already considered something of a modern miracle and widely assumed to be the ultimate solution to the world's energy needs.

Unknown to Thomson or anyone else, though, the answer that solved the riddle of electricity would shortly raise questions that would lead to newer, more exotic, forms of energy with the potential to reduce electricity to relative insignificance.

Up until Thomson's time the atom had always been thought of as a tiny, hard, solid, indestructible speck of matter, usually visualized as a miniature billiard ball. The discovery of the electron altered this concept of the atom forever. Here, embodied in this negatively charged subatomic bit, was a mere fragment of a speck—a small splinter of the atomic billiard ball. The atom wasn't a dab of solid stuff, but an assembly of other, tinier, more basic components. And one of these components—the electron—could be dissociated from its parent atom, pushed through a wire, and made to perform all sorts of wondrous feats. Might there not be other parts of the atom with equal or even greater capacities to provide energy?

That's the question that occurred to Marie Curie when her supervisor, Antoine-Henri Becquerel, showed her how a sample of uranium had exposed a piece of photographic film while in the darkness of a closed drawer. (See Unit I.26.) Curie correctly theorized that the uranium was emitting some sort of energy spontaneously—that energy was somehow pouring out of matter in apparent defiance of the law of the conservation of energy. This energy wasn't the result of a transformation of heat, light, sound, or motion, but something new, different, and as yet totally unfathomed. It seemed to have its source within the unexplored depths of the atom and so, at first, was called atomic energy.

Atomic energy was found to be a spontaneous product of certain heavy elements, the heaviest of which was uranium. These elements released radiant energy—80 percent of it as infrared heat and the remaining 20 percent as deadly gamma rays—and were thus tagged as being radioactive. As the structure of the atom became more familiar it became evident that the energy of radioactive elements came not from the atom as a whole nor from the outlying electrons, but from the nucleus. It was, from that point on, more properly called nuclear energy.

It was Ernest Rutherford who first explained where all this energy was coming from. Using elaborate experiments, indirect evidence, and sound reasoning, as all investigators of the invisible realm of the atom are forced to do, he deduced that radioactive elements are unstable. They can't support themselves in their heavy formats and so gradually crumble into lighter, more comfortable configurations. This deterioration is better known as radioactive decay. As these atoms decompose they release nuclear energy. An atom undergoing radioactive decay, of course, loses its identity in the process. A decayed uranium atom is no longer a uranium atom.

Uranium, as well as most other radioactive elements, ends up eventually as stable lead. This transmutation is a long and complicated process. Uranium, for example, requires 14 separate steps to become lead. Each of these steps is its own special type of radioactive decay. Each involves the rearrangement of

■■

A Famous Formula

The interconversion of mass and energy takes place according to Einstein's deceptively simple formula:

$$E = mc^2.$$

This bit of scientific scrawl has assumed legendary fame, even among those without the slightest idea of what it means. Just for the record: *E* represents energy expressed in ergs, *m* represents mass expressed in grams, and *c* represents the speed of light expressed in centimeters per second.

The speed of light, of course, is a known constant with a value of 30 billion centimeters per second. This makes c^2 (*c* multiplied by itself) 900 billion billion. By plugging this number into the formula it becomes immediately obvious that a little bit of *m* produces one heck of a lot of *E*.

The actual conversion of matter into energy, however, is much less efficient than the mathematics would suggest. Less than 1 percent of the mass of the uranium or plutonium "burned" in a nuclear reactor becomes energy according to Einstein's law. The remaining 99 + percent remains behind as "ashes," the various materials that constitute nuclear waste.

■■

subatomic particles and each generates nuclear energy—heat and gamma-ray radiation. Some of these steps are relatively fast and violent while others are so subtle they're virtually impossible to detect. Each requires time—anywhere from a finger snap to 80,000 years. As uranium decays into lead, then, it assumes a whole series of other identities along the way. This continual and gradual transformation can be tracked and dated by measuring the types and amounts of each of these intermediate substances. The age of moon rocks, meteorites, and the earth's crust can be calculated by comparing the ratio of uranium to its various products in much the same way that the age of once-living fossils can be calculated by using carbon-12/carbon-14 ratios. (See Units I.28 and IV.4.)

Even after the general mechanics involved in radioactive decay were described, though, the specific source of nuclear energy remained unidentified. Finally, in 1932, John Cockcroft (1897–1967) of Great Britain and Ernest Walton (1903–) of Ireland explained the mystery. They succeeded in weighing unstable atoms before they disintegrated and then in weighing the results of the disintegration process. What they discovered was that in every instance the products of radioactive decay weighed ever so slightly less than the predecay atom.

Cockcroft and Walton measured a number of radioactive events and concluded that the amount of energy liberated during the process was directly related to the amount of mass lost according to a mathematical formula. Finally, they determined that this mathematical relationship between lost mass and gained energy was the same one proposed in 1905 by a then-unknown clerk in the Swiss patent office named Albert Einstein (1879–1955). And so, 27 years after he brashly predicted that matter could be transformed into energy, that the law of the conservation of matter—intact since Lavoisier first proposed it in 1774—needed to be revised, that matter and energy were interchangeable aspects of a single phenomenon, Einstein was proved correct. What had been a radical new theory was suddenly a radical new reality.

II.30 A CHAIN REACTION

The vast potential of nuclear energy was immediately obvious. A quick application of Einstein's formula revealed that the amount of energy locked up inside matter was nothing short of limitless. There remained the problem of finding a key to open that lock, however. How could common, abundant matter be transformed into precious energy on command? How could the nucleus be tapped to provide usable amounts of power when and where it was wanted? So far the only known transformation of matter into energy took place during natural radioactive decay, and this process was too rare, slow, and unpredictable to serve any practical purposes.

The decay process, it was then learned, could be speeded up or even induced in nonradioactive substances by artificial means. Manipulating and goading the nucleus resulted in enhancing the amount of energy it yielded. Although early experiments along these lines consumed more energy than they produced, they held out some promise of a breakthrough in the quest for the unequalled prize of unlimited power.

It was in 1938, in frenzied pursuit of this goal, that two German researchers, Otto Hahn (1879–1968) and Fritz Strassmann (1902–1980), fired a stream of neutron "bullets" into a sample of pure uranium. What they had hoped to accomplish with their experiment was the fabrication of new elements that were heavier than uranium and more violently radioactive. The results were disappointing. The bombarded uranium had not become any new, heavier element, but had instead partially reverted into much lighter and familiar elements: barium and krypton to be specific.

This unanticipated and confusing behavior of uranium was explained the following year by Lise Meitner (1878–1968) and her nephew Otto Frisch (1904–1979), two Austrians working in Sweden where they had fled to escape Nazism. The uranium atoms had not absorbed the neutrons fired at them, they said, but had been split asunder by these subatomic bullets. The original uranium nuclei had been split into nuclei of lighter, simpler atoms—those of barium and krypton. This splitting of larger atoms into smaller ones they called nuclear fission. Meitner and Frisch made another discovery as well. A uranium atom undergoing fission produced not only one atom of barium and one atom of krypton, but anywhere from two to three neutron bullets similar to the one that caused the fission in the first place.

This last revelation was of tremendous significance. A uranium atom, split by an initial neutron bullet, released energy and other neutrons that were then capable of splitting other uranium atoms, thereby releasing more energy and more neutrons, and so on. The process was theoretically capable of sustaining itself if enough uranium nuclei were available to split. Not only that, under ideal conditions the process would not only continue unassisted, but rapidly escalate. A single neutron bullet splitting a single uranium nucleus would release two or three other bullets. These two or three bullets would then split two or three other nuclei and release a total of anywhere from four to nine more bullets. These would split more nuclei and generate anywhere from eight to twenty-seven more bullets. This accelerating spiral of cause and effect was the key physicists had sought to unlock the floodgates of nuclear energy. They called it a chain reaction.

With the discovery of the chain reaction the useful production of nuclear energy seemed destined to be achieved. All that was needed to transform it from a theoretical to a practical reality was the appropriate technology. The effort to develop that technology, occurring as it did during the early years of World War II, became a frantic race. The nation that first harnessed the tiny atom, it was understood, would achieve military supremacy and attain global dominance. A war that had started out as a battle among soldiers was to become a battle among scientists and engineers.

In the United States President Franklin Roosevelt responded to the urgings of concerned scientists, with the German-born Einstein acting as spokesman, by launching the Uranium Project. The stated purpose of the project was to study the feasibility of creating a self-sustaining nuclear chain reaction. The project was called the Uranium Project because uranium was the substance that offered the greatest promise of success. Of the 92 natural elements, uranium, number 92, is the heaviest, most complex, least stable, and therefore most vulnerable to artificially induced fission. The size of the uranium atom lies at the very limit of atomic capacity. Heavier atoms have been produced in the laboratory and may have existed in nature at one time, but they all quickly disintegrate into simpler, more durable formats. Lighter atoms can be subjected to nuclear fission, but none produce the extra neutron bullets needed to sustain a chain reaction. Among the natural elements, then, uranium is uniquely qualified as a nuclear fuel.

Not even all uranium is fissionable, however. Uranium exists in several variant forms called isotopes, which exhibit identical chemical behavior but have different atomic weights. (See Unit I.28.) The overwhelming majority of uranium atoms consist of 92 protons and 146 neutrons and have an atomic weight of 238 amu. These atoms comprise what is known as uranium-238, or U-238. This most common uranium isotope is relatively stable, as its mere existence would indicate, and not suitable for fueling a chain reaction. The isotope of uranium that can be split, that releases nuclear energy, and that is needed to support a fission chain reaction consists of 92 protons and 143 neutrons and is known as U-235. Due to its natural instability U-235 exists only to the extent of seven out of every one thousand uranium atoms.

Uranium-235, the rare isotope of nature's heaviest element, is what the Uranium Project needed to accomplish its goal. More specifically it needed this precious material according to very precisely determined standards of quantity and quality. A chain reaction, you see, won't occur unless fissionable material is present in an adequate supply, called a critical mass. Lumps of nuclear fuel smaller than critical mass size won't sustain a chain reaction because there aren't enough targets available for the neutron bullets to strike. Atoms, remember, are mostly empty space. Without a sufficient number present the bullets sail right through all the available targets and fly harmlessly off into space.

Nuclear fuel, besides existing in sufficient quantity, must also be of a certain purity in order to support a chain reaction. A lump of a raw uranium containing only seven fissionable atoms per thousand won't undergo a sponta-neous nuclear reaction because the neutron bullets strike stable U-238 nuclei and remain lodged there instead of striking unstable U-235 nuclei.

By the end of 1942 the Uranium Project had assembled enough pure U-235 to attempt a chain reaction. The exotic fuel was collected in a squash court under Stagg Field, the football stadium at the University of Chicago. There a team of 42 scientists, headed by an Italian defector from Fascism named Enrico Fermi (1901–1954), kept it carefully divided into subcritical amounts with cadmium insulators. On December 2 of that year, under the direction of Fermi, the cadmium insulators were cautiously removed one by one until the reaction became self-sustaining. Enough insulators were left in place to prevent the reaction from accelerating uncontrollably and exploding. Nuclear power, the most awesome power known, had been wrested from nature and placed in the custody of humanity.

II.31 | THE ATOMIC BOMB

Encouraged by Fermi's success, and now deeply entangled in the war, Roose-velt increased the size of the Uranium Project, renamed it the Manhattan Project, and placed it under strict military control. The project's purpose now was to create a functional nuclear explosive device—an atomic bomb—with which to win the war.

To provide the needed U-235 a huge processing plant was built in Oak Ridge, Tennessee, sometimes referred to as Atomic Bomb City. Because isotopes like U-238 and U-235 are chemically identical it was necessary to separate them based on their physical difference—their weight. Raw uranium was first combined with fluorine to produce uranium hexafluoride, a gas. The resulting gas molecules—most containing heavier U-238, a few containing the lighter, precious U-235 atom—were then repeatedly filtered or sifted through a Teflon membrane via a process called gaseous diffusion. Once the lighter gas molecules had been winnowed out they were chemically reduced to yield flourine gas and nearly pure U-235.

Meanwhile it was discovered that U-238, subjected to the conditions inside Fermi's nuclear reactor, transmutated into a new element, plutonium, which was an even better candidate than U-235 for nuclear fission and chain reactions. A new reactor was hurriedly constructed in Hanford, Washington, and the production of plutonium begun.

While quantities of fissionable materials were being painstakingly stockpiled in Tennessee and Washington a group of scientists in New Mexico, headed by J. Robert Oppenheimer (1904–1967), an American, was readying the apparatus to build the world's first atomic bomb. The laboratory itself was situated in Los Alamos, while a site for a test blast was selected near Alamogordo.

As it turned out, plutonium supplies were readied first and a plutonium test bomb assembled. Amid much anxiety the device was successfully detonated at 5:30 A.M. on Monday, July 16, 1945. Three weeks later, on August 6, a U-235 bomb was dropped on Hiroshima, Japan. On August 9 a plutonium bomb was dropped on Nagasaki. The very next day the Japanese opened the negotiations that would result in their surrender and end the war.

The atomic bomb, or A-bomb, whether it be of U-235 or plutonium, works on the same chain-reaction principle as did Fermi's first reactor. The difference is that in the bomb there's no effort made to control the pace of the reaction. It's allowed, or rather encouraged, to run wild. An atomic bomb is triggered by a conventional explosion that rams together several chunks of fissionable material. Each of these chunks is of a subcritical mass. When combined, though, they assume a mass that is critical. The ensuing chain reaction escalates to explosive proportions within a few millionths of a second. It's the speed of a nuclear explosion as much as anything else that provides its great destructive power. The duration of a nuclear explosion compares to that of a dynamite explosion the way a second compares to a century. The speed of the explosion produces a huge sonic boom–like shock wave that accounts for most of the bomb's effect. The bomb also generates an enormous amount of heat as well as lethal doses of gamma radiation. Finally, an atomic bomb creates a cloud of hazardous radioactive remnants, called fallout, which can linger in the atmosphere for years and spread over the face of the globe.

The same forces that leveled Hiroshima and Nagasaki are captive within nuclear generating plants. These reactors are huge, sophisticated versions of Fermi's experimental reactor at the University of Chicago. Either U-235 or, more commonly, plutonium is "burned" in these plants in order to produce heat. The heat is then used to turn water to steam, and the steam is used to drive turbine generators that produce electricity via electromagnetic induction. In some reactors ordinary U-238 is converted into fissionable plutonium while either U-235 or plutonium itself is being used to generate electricity. These so-called breeder reactors actually create more usable fuel than they consume during the course of their operation. This doesn't mean that these plants, or nuclear power in general, has a bottomless well of fuel upon which to draw, however. Even breeder reactors need U-238 to create plutonium, and U-238 exists in very limited quantities.

All nuclear reactors generate electrical power by converting mass into energy according to Einstein's formula. The actual amount of mass annihilated during this process is very small. Uranium or plutonium that undergoes fission retains almost all of its original mass. Even the total number of subatomic particles remains the same during the fission process. Only their arrangement as fewer, heavier atoms into more numerous, lighter atoms changes. These new, smaller atoms, in total, weigh slightly less than did their larger predecessors. Uranium and plutonium are not so much consumed as they are transmutated into other elements.

It's these resultant elements, many of them existing in exotic, radioactive isotopes, that constitute the toxic waste products of nuclear reactors. These wastes, like the fallout from atomic bombs, have long-range and far-reaching potential for wreaking serious and irreparable environmental damage. Once a dangerous, unstable atom has been created it will eventually undergo radioactive decay and release deadly gamma radiation as it does so. Nothing—not diluting it, not gasifying it, not combining it chemically with other atoms, not isolating it—will prevent this from happening. What's more, nothing we can do can dictate or even determine just when this event will take place. Each unstable atom is like a miniature time bomb set to go off at some unknown time. There's no way to defuse it and no way to intentionally detonate it. The best we've been able to do so far with these deadly, uncontrollable, and unpredictable waste products is to stash them as far out of harm's way as possible and hope they decay without causing any damage as they do so.

II.32 | NUCLEAR FUSION

The intensive exploration of the inner workings of the atom that led to the discovery of the chain reaction and the subsequent development of the A-bomb and nuclear reactors uncovered other useful information as well. Most interestingly, the tendency of heavy elements to decompose into medium-weight elements was found to be mirrored by a tendency of extremely light elements to coalesce into medium-weight elements. And, just as the results of heavy-element decay weighed slightly less than their starting materials, so did the results of light-element combination, more properly called nuclear fusion.

Atoms of average weight, it appears, need less energy to hold themselves together than do those at the extremes of the periodic chart. These medium-sized atoms represent the most stable, most desirable, easiest to maintain atomic structure. The extra energy a very heavy or very light atom needs to maintain its somewhat unconventional format becomes unneeded as it assumes a more average

■■■■■■■■■■■■■■■■■■■■■■■■■■■■■■■■■■■■

Solar Energy

Before the discovery of fusion no one had any idea what made the sun shine. The most efficient fuel known on earth was coal, but there was simply no way coal could produce the amount of heat that the sun did. The quantity of fuel consumed by a coal-fired sun burning at its present rate over the past few thousand years would have filled the solar system to a point beyond the earth's orbit. At its current size a sun made out of solid coal would rather shortly run out of fuel. Fusion guarantees that no such calamity will befall us for quite some time. The sun has enough hydrogen fuel for many millions of years. Besides supplying heat, the fusion process also generates new kinds of matter. After hydrogen atoms fuse to form helium atoms the helium atoms will, at high enough temperatures, fuse to form carbon atoms, and so on. This buildup of heavier elements from lighter ones is, we suppose, how the elements were originally created in some primordial fireball. The hottest stars may still be producing a whole array of elements from nothing but simple hydrogen, or possibly from nothing more than subatomic bits and pieces.

■■■■■■■■■■■■■■■■■■■■■■■■■■■■■■■■■■■■

size. This unneeded energy is what is released during either fission or fusion. Energy from these seemingly opposing processes thus has a common source.

Just as the heaviest of atoms—those of uranium—are the best candidates for fission, so the lightest of atoms—those of hydrogen—are the best candidates for fusion. The fusion of hydrogen atoms doesn't rely on neutron bullets or chain reactions, but on simple heat. At temperatures beyond imagining four hydrogen atoms collapse to form a single atom of helium. The helium atom produced weighs fractionally less than did the four hydrogen atoms from which it arose. The lost mass is liberated as pure energy, namely more heat. Once a fusion reaction is ignited it will continue in this manner, much like a conventional fire, as long as there's fusable fuel available.

This, in fact, is what's happening continually on the sun and all the other stars. Every second the sun turns something like 657 million tons of hydrogen into 653 million tons of helium. The remaining 4 million tons of matter are coverted into energy. A tiny fraction of this output—about two parts out of every billion—falls on the earth as radiant energy, either visible light, infrared heat, or ultraviolet radiation. That means that virtually the entire energy needs of the planet for a second are supplied by about the same amount of mass as found in a bucketful of dirt.

It's the fabulous power of fusing hydrogen atoms that ultimately lights our days, warms our planet, grows our food, creates our rain, and makes life on

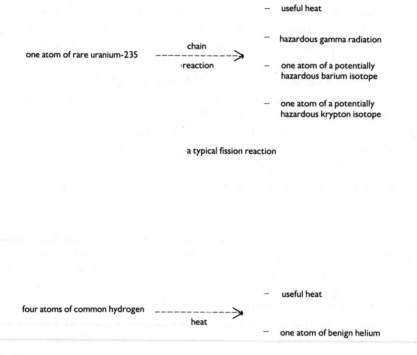

Figure II.5   Fission vs. fusion

Earth possible. It's also the energy of nuclear fusion that has been captured over the years as chemical energy and stored for later use that provides us with all our coal and petroleum reserves. It's been calculated that no less than 99.98 percent of all our energy comes either directly or indirectly from the sun. The only .02 percent that doesn't is energy supplied by the sloshing of the oceans in their beds, the geothermal energy of geysers and volcanoes, and the energy we produce by our own feeble attempts at nuclear fission.

Fusion reactions, also called thermonuclear reactions, generate about two and one-half times more energy than do fission reactions on a per atom of fuel basis. On an atomic weight basis their advantage is much more pronounced, for the weight of hydrogen atoms is only a tiny fraction of that of uranium atoms. Their biggest advantage, though, is the fact that hydrogen, unlike uranium, is plentiful, cheap, and easy to obtain. The hydrogen in a canteen of seawater could easily provide enough fuel to propel the family sedan from Dad's New York apartment to Mom's Malibu beach house if used in a fusion reaction. Finally, fusion is a clean process, meaning that it doesn't produce any radioactive waste the way fission does. The only by-product of fusion is helium, a benign, inert gas. Fusion—efficient, limitless, and harmless—would indeed seem to be the final answer to our energy needs. (See Figure II.5.)

Unfortunately, the management of a fusion reaction hasn't yet been mastered. The only human-made fusion reactions produced so far have been violent ones: hydrogen bombs, also called H-bombs, fusion bombs, or thermonuclear bombs. The temperatures needed to trigger a thermonuclear reaction are so high that they've so far only been achieved with a nuclear fission explosion. H-bombs, like the first one exploded by the United States on Eniwetok Island on November 1, 1957, are detonated by A-bombs. The hydrogen bomb, unlike the atomic bomb, can be built without any limit to its size. The first H-bomb was 750 times as powerful as the bomb that destroyed Hiroshima, and much, much bigger ones have been constructed since. Thousands of these miniature suns lie menacingly awaiting detonation.

Fusion remains uncontrollable because of the extreme temperatures it involves. To be useful a fusion reaction must be contained while its enormous output of heat is channeled into desirable outlets. But no physical container can withstand such heat. It's like trying to find a bottle in which to store a universal solvent. But nuclear physicists, ever imaginative, haven't conceded defeat. They're attempting to build a "bottle" made not out of any material substance but out of powerful magnetic fields. The hope is that some day, probably with the aid of lasers, a fusion reaction will be triggered and suspended in midair.

If the past is any indication of the present, they'll eventually succeed. When they do, nuclear fission will be as old-fashioned as kerosene lanterns are now. Fusion, in turn, could itself be supplanted by an as yet undreamed-of form of energy. But whatever forms it might take, the ultimate source of energy, like the ultimate nature of matter, would seem to lie deep inside the atom, where energy and matter reside together in a paradoxical state of unity and duality.

III
SPACE

"Astronomy compels the soul to look upward and leads us from this world to another."

—Plato, c.428–c.347 B.C.

III.1 THE ENDLESS EXPANSE OF SPACE

From the very earliest glimmer of human consciousness it was the mastery of matter and energy that assumed paramount importance. Giving a crystalline solid spear point enough momentum to pierce the hide of a predator could mean the difference between surviving and perishing. Hacking a cave out of a hillside of chemical compounds and warming that cave via the rapid oxidation of hydrocarbons could mean the difference between comfort and misery. But it was the endless expanse of space that first captured the human imagination. Spending night after night outdoors—without television, radio, reading material, or even very many neighbors to talk about—our ancestors passed countless hours gazing at a pristine panorama of stars and wondering, as we still do now, just what it could be.

For thousands of years this combination of scrutiny and curiosity did little to make the sky any less bewildering. Failing utterly to understand it in any familiar terms or to find in it any earthly purpose, the ancients assumed the sky was something fantastic, supernatural, and mysterious. To the Egyptians it was the starry body of a goddess arched over the earth like a protective umbrella. To the Babylonians it was a huge bell-shaped dome, speckled with pinpricks of light. To the Arabs it was a vast tent of black velvet, the inside of which was sewn with white stars. The specific imagery varied from culture to culture, but the basic concept was much the same—the earth was a relatively small, flat, stationary disk over which wheeled a distant, star-spangled ceiling of some sort.

There were a few things that couldn't be explained by this simple, two-part scheme of a central, static earth and a far-flung, mobile canopy to which the stars were affixed, however, Specifically, the activities of seven heavenly bodies corresponded with neither the resting position of the earth nor the nightly rotation of the nocturnal vault. First and foremost there was the sun. Invisible at night, it apparently had nothing whatsoever to do with the stars. And during the day, when it blazed with unequaled brightness, it obviously moved quite independently of the earth. Then there was the moon. It was visible sometimes at night with the stars, sometimes during the day with the sun, and sometimes not at all. Even its shape changed during the course of its confusing travels. Finally, out of the countless thousands of stars, there were five very special ones, identified now as planets, which roved relentlessly and irregularly against the unchanging backdrop of their nighttime peers.

While the sky in general continued to awe and humble anyone giving it anything more than a casual glance, these seven objects in particular became the subjects of special speculation, wonder, and even worship. Each was assigned its own domain, one that was in each case assumed to lie somewhere between the innermost domain of the earth and the outermost domain of the regular stars. Each of these seven bodies was endowed with its own identity, free will,

● ●

The Planets

The five planets readily visible to the naked eye and known to all ancient stargazers are: Mercury, Venus, Mars, Jupiter, and Saturn. Venus, the third-brightest object in the sky after the sun and the moon, is hard to miss. Mars and Jupiter are also brilliant, with Mars having the added distinction of appearing red. Saturn is rather ordinary looking, but presents no challenge to the serious observer. Spotting Mercury, though, is a bit tricky. Because of its nearness to the sun it never appears when the sky is completely dark.

This list remained unaltered from prehistory until 1543. It was then that Copernicus added a sixth name by announcing that the earth itself was a planet and not the immovable center of the universe as previously supposed. (See Unit III.4.) The seventh planet, Uranus, wasn't discovered until 1781, and then only by accident. Uranus, although faint, can also be seen with the unaided eye under good conditions. The ancients probably saw it but, because its motion is so slow, failed to identify it as a planet.

Neptune, the next planet to be detected, was identified in 1846. Although it can be seen with a pair of binoculars, it was actually discovered with pencil and paper before it was ever seen. An irregularity in the orbit of Uranus suggested the gravitational tug of another planet. Tipped off about where to look by a mathematician, it took astronomers less than a year to find it. Pluto, the ninth planet, was found the same way in 1931. Some evidence suggests that other planets, undoubtedly very small and very distant, may also exist.

—Our word *planet* comes from a Greek word meaning wanderer. Can you identify Venus and Mars in the nighttime sky? Track them for a few hours, or even better, for a few nights. Do their positions change with respect to the stars?

—When astrologers claim they can foretell the future based on the stars they are usually referring to the planets. Why are the planets a better source of astrological information than true stars?

—Mercury and Venus, the planets between the earth and the sun, exhibit phases in much the same way the moon does as it waxes and wanes. Can you explain why?

—Venus takes about seven and a half months to orbit the sun, Mars close to two years, and Jupiter almost twelve years. Why do they appear brighter at some times than at others?

● ●

spirit, intelligence, and power. Eventually, the seven errant travelers assumed the guises of seven heavenly beings. They became gods.

The elevation of the sun, the moon, and the five visible planets to a position of superiority necessarily reduced the status of the earth and its inhabitants to inferiority. The sacred seven could be admired, but they could never be understood. The human intellect, being of an inferior realm, was by definition a powerless and pitiful tool with which to probe the heavens. The stars were beyond its reach.

III.2 LOGIC AND GEOMETRY

The sense of helpless resignation regarding the heavens, the feeling that they were off-limits to human inquiry, persisted until the rise of Greece. The Greeks believed in gods and goddesses all right, and had plenty of them, but they also believed in their own mental abilities and the power of rational thought. They saw no reason why the human mind should timidly shrink away from the exploration of space. To them heaven was simply an extension of the earth, not a separate reality. It was a distant, untouchable extention, sure enough, but it was nonetheless considered to operate according to laws of some sort, and these laws were considered to be intelligible to the human mind. The stars, like matter, beauty, and virtually everything else, were fair game for Greek curiosity.

The first accomplishment stemming from this cocky new attitude occurred on a day we now reckon as May 28, 585 B.C. On that date Thales of Miletus successfully predicted a solar eclipse. This was no doubt not the first time such an event had been accurately foretold. The Egyptians, the Babylonians, and several American Indian cultures had previously attained sufficient proficiency in astronomical record keeping to enable them to perform such a feat. It was, however, probably the first time such an event had been received with anything besides terror, wonder, panic, and expectations of disaster and doom.

Thales, unlike his predecessors, didn't believe that the sun god was dying, that the sun's fire was being extinguished by the forces of evil, or that the end of the world was at hand. No. Brashly using confidence, logic, and geometry, Thales explained the eclipse in physical rather than mystical terms. The sun circled the earth, he reasoned, as did the moon. The sun's round-trip took exactly one day; the moon's slightly longer. The sun, with its marginally faster orbital speed, thus periodically overtook the slower moon. According to Thales, the moon's orbit was smaller and closer to Earth than was that of the sun. When the sun overtook the moon, then, it did so on the outside, so to speak. The moon resided between the earth and the sun at these times and the nearer moon blocked out the farther sun. This, not black magic or angry angels, was what caused a solar eclipse. (See Unit III.14.)

Other Greeks as well, lured by the thrill of solving the ultimate brainteaser, focused their considerable mental facilities on the mechanics of the universe. There was, for example, Hipparchus (2nd century B.C.), who properly interpreted a lunar eclipse. When the orbiting sun and moon were diametrically opposite one another with the earth in between, he said, the earth threw its long shadow on the moon, thereby darkening it. (See Unit III.14.) Hipparchus also measured the time it took this shadow to move across the lunar face, threw in a bit of trigonometry, and calculated the distance from the earth to the moon with an astounding 99.7 percent accuracy.

The idea that the sun and the moon orbited earth in regular, if complicated, fashions not only satisfactorily explained solar and lunar eclipses, it implied that the earth was round, not flat. Thales himself declared that "the earth is a globe and inhabited on all sides," a progressive thought to say the least, and one that had probably never so much as flickered across a non-Greek mind. Thales reached this conclusion during a journey from his native Miletus to Egypt, much farther to the south. There he noticed that the Great Bear constellation, which was always visible farther north, sometimes dipped below the horizon. He realized that by traveling south he had changed his vantage point on the sky. The difference in perspectives at Miletus and Egypt could nicely be explained by assuming that he had moved over the surface of a sphere during his trip.

Again, it was a powerful demonstration of the ability to defy convention, discard common knowledge, and think originally and abstractly. Thales' conclusion seemed to have some merit. Not only did a round planet explain the various views of the sky, it also agreed with the shape of the darkness sweeping across the moon during a lunar eclipse. Maybe it really was the earth's shadow. And, closer to home, a spherical world also explained why a ship sailing out to sea gradually sank below the horizon regardless of which direction it took.

Eratosthenes (c.276–c.192 B.C.) not only figured that the earth was round, but calculated its size. This he did by measuring the angle of the sun's rays incident on the earth's surface first at one location and then at another, relatively distant, location, each time doing so at high noon. Starting only with knowledge of the two angles, the distance between his two observation points, and a few trigonometrical techniques, Eratosthenes came up with a circumference of the globe that substantially agrees with that computed by present techniques.

To the Greeks, then, the earth was simply a large sphere suspended in three-dimensional space. The sun, moon, and planets were inanimate objects rather than conscious beings. The universe beyond was a physical rather than a spiritual place.

III.3 THE GREEK UNIVERSE

Daring, brilliant, and ambitious, the Greeks enjoyed a thoroughly impressive amount of success in deciphering and explaining the relative sizes, positions, and motions of the earth, the sun, and the moon. Then, under the personal supervision of Aristotle, these gains were consolidated. The result was an intellectual model of the universe that Aristotle felt explained everything in need of explanation.

According to this model the universe consisted of two distinct parts: the earth and distant heavens. Each part was composed of different kinds of stuff and operated according to different laws. The immediately accessible portion of the universe was a conglomeration of four elements: earth, water, fire, and air. The portion that was physically out of reach, on the other hand, was an unsullied ocean of a fifth element called aether. Terrestial motion was linear, finite, and disjointed, consisting of vertical and horizontal starts and stops. Celestial motion, in contrast, was curved, constant, and eternal.

At the center of this universe was a round earth. Surrounding it were the sun, the moon, and the five vagabond stars, each of which was mounted on an unblemished, transparent, and mobile crystalline sphere. These seven flawless spheres were arranged concentrically about the earth like the layers of an onion. The moon occupied the closest layer, Saturn the farthest. Beyond Saturn was yet another eighth and ultimate sphere to which were attached the regular, or fixed, stars.

Because celestial motion was curved and constant, all heavenly bodies necessarily traced out perfect circles. The motion of both the sun and the moon nicely fulfilled this requirement. The appearance of the fixed stars also made sense if one assumed they were mounted on a rotating sphere. But explaining the paths of the remaining five members of the sacred seven in terms of perfect earth-centered circles proved to be a bit more difficult.

To begin with, their distances couldn't be properly estimated, for they were all too far away to be compared to anything measurable. They might be twice as far away as the moon or 20 times as far away, nobody could tell. The irregularity of their routes and the fact that they periodically got brighter and dimmer even suggested that they might not always remain at a constant distance. There was, in fact, no earth-centered circle, nor any other simple geometric description, that explained the path of any of the five wandering stars.

It was the planets, then, those five oddballs of outer space, that threatened to unravel Aristotle's neatly wound ball of yarn. Someway, somehow, somebody had to figure out how to construct a model of the universe that not only faithfully described the motions of the planets, but also satisfied Aristotle's philosophical requirements.

Easier said than done. Totally stymied for a long time, Greek mathematicians finally came up with a geometric gimmick called an epicycle, a complex construction of multiple, interacting circles. According to this contrivance the planets indeed moved in perfect circles, but the focal point of each planetary orbit wasn't the earth but an imaginary point that moved in a circle about the earth. The sun, the moon, and the cosmic perimeter, with no epicycles, orbited the earth like horses about the axis of a merry-go-round. A planet with an epicycle, meanwhile, orbited the earth like a horse on merry-go-round that was in turn mounted on the edge of a second merry-go-round. By fiddling with the sizes and speeds of the involved carousels, all sorts of looping motions could be assigned to the planets, some of which fairly well coincided with the observed facts.

As observational data accumulated, however, it became increasingly apparent that the model provided only a crude approximation of actual planetary motion. Its shortcomings were significant and had to be remedied. Unwilling to admit they had erred, unable to abandon Aristotle's hallowed circles, the Greek astronomers did the only thing they could do—they refined their model. They added more epicycles. Circles sprouted upon circles. Imaginary points and merry-go-rounds multiplied. The resulting monstrosity was extremely cumbersome and the calculations involved in describing the path of a planet excruciatingly tedious, but circlar motion was maintained.

During the remaining years of Greek supremacy the Aristotelian worldview was enlarged upon, refined, and established beyond a reasonable doubt by his faithful followers. Finally, around 150 A.D., Claudius Ptolemy of Alexandria (2nd century A.D.) collected the works of all the astronomers who had preceded him and set down what was to be the final version of the Greek universe. It was, for the most part, Aristotelian. It was, for example, geocentric, meaning that at its center—motionless, imperial, and preeminent—resided the earth. Around this heavenly focal point, at various distances, revolved the moon, the sun, the five planets and, finally, the celestial dome.

Ptolemy, ever faithful to Aristotle, employed a total of 39 epicycles in order to maintain uniform and circular motion in the heavens. With this many merry-go-rounds he felt satisfied that his pencil-and-paper universe reproduced the movements of the physical universe with sufficient accuracy. Whether or not one had anything to do with the other, however, was quite another thing altogether. But it didn't really matter, for geometry was as close to the stars as the Greeks expected to get.

Ptolemy published his arrangement, along with tables, charts, and diagrams, in a 13-volume work he titled *Mathematike Syntaxis*. This work would eventually come to be known as the *Almagest,* a name of hybrid Greek-Arabic origin usually translated simply as "The Greatest." For the next thousand years and beyond the *Almagest* would serve as the unquestioned authority on cosmic matters and Ptolemy, for better or worse, would be to astronomy what Aristotle would be to chemistry and physics.

III.4 A REVOLUTIONARY IDEA

Ptolemy ruled as high priest of the heavens for over a millennium for three main reasons. In the first place, there just weren't any other serious contenders for his throne. With the Greek demise curiosity about the material world became all but nonexistent. The knowledge acquired by the Greeks was casually accepted as the ultimate truth by those who followed them. There was only so much to be learned, and the Greeks had learned it.

Secondly, whatever scant academic or scientific activity did exist supported rather than challenged Ptolemy's position. Nearly all such activity took place within the structure of the Catholic Church, which, filling a void left by the fall of the Roman Empire, served as an authority on intellectual, governmental, social, economic, and even military matters as well as religious ones. And the Church enthusiastically endorsed Ptolemy, much as it did Aristotle. Not only did the Church find his scheme astronomically workable, it found it theologically palatable as well. It was altogether proper that the earth, the crowning glory of God's creation, should occupy an exalted central and permanent position.

Finally, there was good old common sense. Anyone with eyes to see with knew that the sun, the moon, the planets, and the fixed stars revolved around an undeniably solid and stationary earth. Nothing could be more obvious or less open to discussion. Why waste any more time and energy on something as remote and irrelevant as the stars?

But, as the centuries passed, interest in the heavens grew. This interest wasn't driven so much by scientific or intellectual motives as by practical ones. In the Middle East the Muslims needed to know the proper times for their prayers and the proper direction of Mecca. In the seagoing countries ever more adventurous traders, explorers, and colonizers needed navigational aids to find their way to uncharted regions and safely back home again. And all across Europe astrologers were peering into the future, telling fortunes, diagnosing diseases, and prescribing medicines for their cure. In every case it was the sky that was consulted for guidance in these matters.

With the renewed scrutiny of the stars came frequent and rigorous tests of Ptolemy's universe. The results weren't very favorable. In almost every instance the astronomical positions and motions actually observed differed from those projected by Ptolemy's charts and tables. At first these discrepancies were rather small and easily attributed to minor errors on the part of either the observers or the interpreters of the *Almagest*. But as time passed the irregularities continued to grow both in number and in magnitude. Like clocks running at slightly different speeds, the real world and Ptolemy's world were getting farther and farther out of synch. The stars and planets weren't where they were supposed to be. The tides didn't ebb and flow on schedule. Even the seasons of the year seemed to be drifting across the calendar.

This was the situation when Nicholas Koppernigk (1473–1543) was born in Poland. Koppernigk, who later called himself Copernicus, became quite familiar with these problems during his education both as a churchman and as a scholar. After studying theology, philosophy, law, and medicine, and mastering Greek and Latin, he was one of the best-educated men of his time. The schism between how the sky looked and how it was supposed to look troubled him immensely, and he dedicated much of his life to its resolution.

Copernicus reasoned, simply enough, that either Ptolemy's model or the observed facts must be in error. As a scientist and independent thinker he chose to believe his own eyes rather than the ancient texts. But, as a devout Aristotelian, he was unable to abandon Ptolemy completely. So he took a middle course and set out to see if he couldn't improve and refine the basic doctrine of the *Almagest* by adding, subtracting, or altering a few epicyles and recomputing its tables.

After years of researching old manuscripts, comparing various star charts, collecting a bit of original data, and laboriously computing orbits Copernicus achieved the geometrically more satisfying universe he sought. It was fairly traditional in that it consisted of perfectly circular celestial motion, constant celestial speed, and a messy array of epicycles. Its superiority in explaining the motions of the planets really rested upon just one alteration of conventional astronomy. Instead of assuming a central and stationary earth it assumed an earth that rotated on its axis once a day and revolved around the sun once a year. The earth, according to Copernicus, was just another planet and the sun, not the earth, was at the center of things. The universe, he said, was sun-centered or heliocentric, not geocentric.

Suggesting that the sun stood still and the earth moved, rather than vice versa, was, of course, a revolutionary idea to say the least, one that contradicted the entirety of existing ecclesiastical and academic opinion, not to mention all the obvious evidence. Copernicus himself was a bit stunned by what he had created and wasn't too sure he was capable of weathering the inevitable storm of controversy and ridicule that would certainly follow its disclosure. And so, paralyzed by fear, he kept it to himself for some 30 years, wondering all the while just what to do with it.

Eventually Copernicus divulged his radical revision to a few close friends, who urged him to share it with the rest of the world. He resisted them at first, but finally, at the age of 69, he succumbed. It was in 1543, as an old and dying man, that he published his now famous book, *De Revolutionibus Orbium Coelestium,* or *Revolutions of the Heavenly Bodies.* As the story goes, it was on his deathbed, safe from repercussions, that he first laid eyes on the printed and bound volume that represented so much of his life's work.

As it turned out, Copernicus could have spared himself all those years of anguish and the last-minute dramatics. His monumental, supposedly outrageous book was greeted not with a torrent of protest, but with what bordered on indifference. The majority of the population didn't care much one way or another what orbited what. The Church, meanwhile, dismissed the book as a mere academic exercise. And common sense still argued overwhelmingly in favor of a stationary earth. All but unreadable, *Revolutions* was anything but a best-seller.

There were a few, though, who found the book to be much more than an intellectual curiosity. Navigators, astrologers, timekeepers, and other users of the sky soon recognized the fact that the charts and tables of Copernicus were

■■■■■■■■■■■■■■■■■■■■■■■■■■■■■■■■■■■

Truly Revolutionary

Since its publication Copernicus's book has attained a symbolic status reserved for only the truly great and revolutionary achievements of the mind. It's from its use in the title of this book, in fact, that the word *revolution* has come to signify a complete and drastic change of some sort.

Curiously, though, the idea of a heliocentric universe had been around for a long time. Aristarchus of Samos first proposed a moving earth some 300 years before the birth of Jesus Christ, and it was references to him in ancient texts, Copernicus admitted, that inspired his revolutionary work.

Still, Copernicus had good reason to be anxious about divulging his revamped universe. Aristarchus had been threatened with indictment for impiety and then banished to obscurity. Copernicus, meanwhile, had to scramble to find someone willing to print *Revolutions*. When it finally did appear, Copernicus discovered that the printer had taken the precaution of inserting a preface that claimed the book was actually just a device to ease calculations and not intended to suggest that the earth really did move.

■■■■■■■■■■■■■■■■■■■■■■■■■■■■■■■■■■■

better than those of Ptolemy and wasted no time in putting them to use. It was its mundane collection of information, then, and not its startling theoretical aspects, that earned *Revolutions* a growing popularity. By this process, backward though it might seem, the book became implanted in the general consciousness. Ever so slowly the idea of a geocentric universe became subject to more and more doubt as the idea of a heliocentric universe became less and less foreign.

III.5 A SIMPLE, CONSISTENT DESIGN

Although putting the earth in motion around the sun as a planet was a huge step in the right direction, the Copernican system still left something to be desired. Despite the improvements it provided there remained a stubbornly persistent gap between theory and reality, and this gap cast a shadow of doubt over Copernicus's efforts in much the same way it had over those of Ptolemy. Astronomical truth was nearer than before, but still beyond reach. There remained much work to be done.

The first priority was to accurately document just how, in fact, the planets really moved across the sky. Existing data had been collected over a span of hundreds of years, by a great variety of observers, from different locations, under different conditions, using different instruments or sometimes none at all. In order to fairly evaluate Ptolemy, Copernicus, or anyone else the standard by which they were to be judged needed to be precisely defined. It was to this task that Tycho Brahe (1546–1601), an astronomer born in a part of Sweden that now belongs to Denmark, dedicated his life.

With financial help from the king of Denmark Tycho built the most elaborate observatory the world had ever seen on an island in the Danish Sound. The station was equipped with sextants and quadrants of huge dimensions, machined to the closest tolerances possible. The observatory itself was constructed partially underground to eliminate distortions that might otherwise arise from vibrations due to wind. There were even correction charts to account for the expansion and shrinkage of the brass instruments due to temperature changes. Here, in this technical marvel of stargazing, Tycho, along with an able staff of assistants, labored for over 20 years collecting the most reliable, most comprehensive, most voluminous, and most accurate astronomical observations ever made.

The job of comparing the various models against this newly affirmed reality was left to one of Tycho's assistants, Johannes Kepler (1571–1630), a German. Kepler was a man with great patience and a passion for mathematics—the perfect choice for the task. He was also a devout Copernican, and so chose to first test the Copernican model. Bursting with enthusiasm and anticipation he plunged into his work.

He started with the planet Mars. His object was this: to plot the orbit of Mars first according to Tycho's data and then according to Copernicus's model, compare the results, determine the extent of the differences, explain those differences, and then make whatever adjustments were needed in the model to eliminate those differences. He bet his colleagues it would take him eight days.

Six years and nine hundred pages of calculations later Kepler was still struggling with the orbit of Mars. Plotting its path first according to Tycho's data and then according to the Copernican model proved to be easy enough. Comparing these two courses and determining the differences between them was also a pretty straightforward assignment. But explaining those differences was impossible. They fit no pattern and gave no clues as to their source. Adjusting the model, then, became a project of pure trial and error. Like a human computer, he first enlarged, then shrank each of Copernicus's circles. He sped them up then slowed them down. He eliminated some epicycles, then added others. After each bit of fiddling he would recompute the entire orbit of Mars and once again, with ever-diminishing expectations of success, compare it to Tycho's data. But try as he might, he could find no combination of circles, speeds, epicycles, and imaginary points that corresponded to reality.

Finally, after having exhausted all reasonable possibilities, Kepler was forced to take a drastic step. He abandoned circular motion as the mode of heavenly operation. Just as Copernicus, trying to salvage Ptolemy, had put the earth in motion, so Kepler, trying to salvage Copernicus, had shattered the

Aristotelian notion of heavenly perfection. The dismantling of the Greek universe was now complete.

But Kepler still didn't know what the path of Mars really was. What followed were more years hunched over paper and pencil, more trials and errors, more computations, and more frustrations. At last, after having experimented with all sorts of imperfect circles, he tried the ellipse, a sort of oval with slightly pointed ends. By using a single ellipse he succeeded in plotting the orbit of Mars around the sun in exact accordance with Tycho's sightings. Trembling with excitement, he tried the same strategy for Mercury, Venus, Jupiter, and Saturn. They all fit! Then, working backward, he plotted the movements of the sun as they would appear from an earth in an elliptical orbit. Methodically, obediently—almost willingly—the facts fell into perfect harmony with his formulas. "I thought at first I was dreaming," Kepler later wrote of this climax to his 17 years of labor.

Kepler published a detailed account of the elliptical orbits of the planets—including the earth—about the sun in 1609 under the title *Astronomia Nova*, or *New Astronomy*. Gone were the random and unwieldy circles circling circles circling circles. In their place was a simple, consistent design of great accuracy.

III.6 A SPOKESMAN FOR THE NEW ORDER

Kepler's triumph by no means marked the end of the geocentric versus heliocentric universe debate. The ellipse, although backed by mathematics rather than philosophy, was in some ways just as arbitrary as the circle. There was, in fact, no real proof that the earth budged, and the Church, tradition, and common sense remained mightily entrenched as advocates of a stationary earth.

The leading advocate of a mobile globe, on the other hand, was none other than Galileo. Although educated in Catholic schools, Galileo was above all a man of science. Regardless of what the scriptures, the Pope, or conventional wisdom told him, he felt the Copernican model of the universe made more sense. At first he kept his heretical views a rather private affair. But in 1610, while on a trip to Venice, he heard about a wonderful new device called a telescope. He returned home to design and build a series of telescopes for himself and for sale, each more powerful than the last. While others used these instruments to identify distant ships or simply as toys, Galileo pointed his toward the heavens. What he saw transformed him from a closet Copernican into a flaming revolutionary and a spokesman for the new order.

The moon, he saw, was a rugged, barren landscape of mountains, rocks, canyons, and craters. The sun was equally blotched with sunspots and solar flares. The heavenly bodies weren't perfectly fashioned, but scarred chunks of common matter. He also saw that Venus, like the moon, waxed and waned as it appeared in various phases. This could only be explained by attributing to it a solar orbit. Finally, focusing on Jupiter, he saw a planet surrounded by its own collection of moons. Venus orbited the sun. The moons of Jupiter orbited Jupiter. Here at last was solid evidence supporting the opinion that not everything had to revolve around the earth.

Galileo was especially struck by the image of Jupiter and its four visible moons. This to him was a god's-eye view of the solar system as he envisioned it—Jupiter representing the sun and its moons representing the planets—and overpowered any reservations he might previously have felt. The moons of Jupiter also demonstrated that a body like Jupiter could be orbited by other bodies even as it was in motion around something else. This observation crushed a popular objection to a moving earth, namely that if it circled the sun it would leave its orbiting moon trailing behind.

With the fervor of a new convert, Galileo began a crusade to popularize the heliocentric universe with a boldness that would have caused Copernicus to shrink in terror. By 1616 the Church had had enough. It commanded Galileo to cease his campaign for the Copernican theory and placed *Revolutions* on the *Index Librorum Expurgatorus*—a list of books Catholics were forbidden to read without special permission. Galileo was reminded that scripture, not science, was the path to enlightenment.

A devout Catholic, and no fool, Galileo restrained himself from further proselytizing but held firmly to his convictions. Then, in 1632, he hatched a new strategy for propagating his views without, he thought, violating the command of the Church. He published in that year his *Dialogue Concerning the Two Chief World Systems—Ptolemaic and Copernican,* which supposedly didn't really advocate the universe of Copernicus/Kepler/Galileo, but merely compared its merits to those of the universe of Aristotle/Ptolemy/Pope Urban VIII. The book, entertainingly written in common Italian rather than in scholarly Latin to ensure maximum readership, was presented as a debate between a Copernican and an Aristotelian. It was heavily weighted toward the heliocentric position, but, according to its author, carefully refrained from drawing any definite conclusions.

The Church was of a different opinion. It felt, with good reason, that Galileo was not only preaching the Copernican doctrine, but mocking its authority as well. On top of everything else, the Pope became convinced that Galileo's fictional spokesman for the geocentric universe, named Simplicio, was a caricature of none other than His Holiness himself. Galileo was summoned to appear before the Roman Inquisition the next year. There he was ordered to recant his views and condemned to a sort of house arrest for the remainder of his days.

Galileo died in 1642, blind, captive, feeble, and no doubt dismayed over the fact that the battle he had so gallantly waged against ignorance, fear, and intellectual tyranny might ultimately be lost. He needn't have worried, though. In Protestant England Mrs. Newton would soon be giving birth to a son and naming him Isaac.

• •

The Trial of Galileo

The trial of Galileo before the Holy Roman and Universal Inquisition in April 1633 was the climactic moment of the confrontation between the old order and the new.

The prosecution was the Roman Catholic Church, trying to hang onto what power it could in the face of the Protestant Reformation sweeping much of Europe. It had recently launched a Counter-Reformation of its own headed by the office for the propagation of the true faith, from whose official title the world *propaganda* is derived.

Galileo, on the other hand, was emboldened by the recent appointment of an acquaintance and admirer of his as Pope Urban VIII. The fact that one of his countrymen, Giordano Bruno (1548–1600), had already been burned at the stake for his heretical astronomical speculations failed to deter him.

The trial hinged upon what, exactly, had been Galileo's instructions when he was first reprimanded in 1616. The defendant claimed that he had been told to neither "defend" nor "hold" the heliocentric view, and had a signed document to this effect. The Church, though, maintained that he was additionally forbidden to "teach in any way whatsoever" the radical ideas of Copernicus. To back up its claim the prosecution produced its own document. This piece of paper remains filed away in the Vatican Secret Archives. Signed by neither Galileo nor any witness, it is commonly regarded as being bogus.

It hardly matters, for the decision of the ten-member tribunal, which included both a brother and a nephew of the Pope, was a foregone conclusion. In the end it all came down to a question of authority, not astronomy, and the Church was not about to decide in anyone's favor but its own.

It's commonly understood, though, that, despite his public apology, Galileo's private convictions remained unshaken. There is even a popular notion that a convicted but defiant Galileo, while rising from his knees before the inquisitor, mumbled under his breath the words *Eppur se muove*—"nevertheless, it (the earth) still moves."

—The Church won the battle. Who won the war?

—Galileo believed in natural laws; the Pope believed in miracles. Who do you agree with? Are the two positions totally incompatible, or could they both be partially correct?

—Can you think of any other clashes between science and religion? Are the two still at odds? Will they remain that way?

• •

III.7 | THE FINAL TRIUMPH

Step by torturous step the heavens were being demystified. The Greeks had transformed them from a purely spiritual realm, populated by lost souls and deities, to a physical one, full of ideal objects moving in flawless circles about a stable earth. Copernicus, trying to make this design more workable, had been forced to put the earth in motion and remove it from the center of the universe. Kepler, with his reams of calculations, then replaced the perfection of circles with the reality of the ellipse. Finally, it was Galileo with his telescope who saw that the sun, the moon, and the planets were in no way pristine, but evidently constructed of much the same sort of stuff found closer to home.

But even though the overall design was finally being understood, there remained a need to explain how the whole thing operated. In times of antiquity it had been thought that the heavenly bodies traveled by means of self-propulsion and free will. The Greeks had relied on whirlpools of distant aether to move the stars. Copernicus pictured the universe as a sort of great clockwork full of phantom gears, pulleys, and flywheels endlessly spinning and driving one another with mechanical precision. Kepler thought that the sun rotated and somehow swept the planets along on giant arms of magnetism that radiated from its center like the spokes of a wheel.

None of these proposed models was even close to satisfactory because they all perpetuated the assumption that the universe "out there" is somehow different from what it is "down here." Each was a fantastic construction of one sort or another that had no worldly counterpart and far exceeded any common experience. Belief in any of them required an act of faith. Because not one of these pies-in-the-sky was supported by even a shred of physical evidence, one could be argued against another endlessly and inconclusively.

Then along came Newton. It was Newton who explained the cosmos in common terrestrial terms rather than in special celestial ones. It was he who welded the physics of everyday life with the astronomy of the stars to form a truly unified universe governed by a single set of all-purpose laws—a universe of such logic, such integrity, and such elegance that it possessed an almost undeniable validity. It was Newton who led the revolutionary forces to their final triumph.

Newton achieved this victory because he was able to appeal to reason. Common sense, which argued against all his predecessors, was Newton's ally. Newton's universe required no leap of faith, no imagination, no fantasy, and no invention, for it relied on something literally as simple as falling off a log.

Gravity. Common, everyday, utterly reliable, and familiar gravity explained the shape of the heavenly bodies, their motions, and the configuration of the solar system.

Newton didn't discover or invent gravity, of course, but he was the first to properly grasp its significance. His interest in this utterly ordinary yet highly extraordinary business was piqued in 1665 while he was spending time at his country home in Lincolnshire to escape the plague. It was during this period of isolation, away from all distractions, that Newton was free to focus the full power of his incredible mind on the mechanical and mathematical abstractions that so enthralled him. Passing the days in thought and calculation, he discovered the binomial theorem and the method of infinite series, developed differential and integral calculus, built telescopes, conducted his famous experiments with light and prisms and, just for the fun of it, computed the area under a hyperbolic curve to 52 decimal places. As legend has it, it was the sight of an apple dropping from a tree that turned his attention to gravity.

Where others would have seen a falling piece of fruit Newton saw a principle of physics. The apple fell because it was attracted to the earth, that much was obvious. But, theorized Newton, the earth was also attracted to the apple. The matter content, or mass, of the earth and the matter content, or mass, of the apple, exerted a strange drawing power on all other matter. All matter was mutually attracted through the power of gravity.

With this principle in mind Newton went back to Galileo's experiments with falling bodies and inertia, described the results using his newly created calculus, and came up with his three laws of motion. (See Unit II.6.) Not content with having achieved a major breakthrough in terrestrial mechanics, he then focused his attention on the heavens. Doing to Kepler what he had done to Galileo, Newton, again with the aid of calculus, described elliptical planetary motion with a mathematical formula.

Having described both terrestrial and celestial mechanics with mathematical precision, he was ready for the crowning touch. He was ready to bridge the gap between them. Using his laws of mechanics and falling bodies on the one hand and his summary of Kepler on the other, he "compared the force requisite to keep the Moon in her Orb with the force of gravity at the surface of the Earth, and found them to answer pretty nearly." Both events—the terrestrial descent of the apple from its tree and the celestial circuit of the moon—could be accurately and equally described by the same set of equations. The universe, arbitrarily split into upper and lower realities for thousands of years, was at last whole.

III.8 UNIVERSAL GRAVITATION

When Newton found the fall of the apple in his orchard and the orbit of the moon about the earth "to answer pretty nearly" according to his laws of universal gravitation and motion he didn't run through the streets shouting "Eureka!" as anyone else who had just discovered the secret mechanism of the universe might be expected to do. No, not Newton. The reclusive bachelor, having solved a challenging riddle to his own satisfaction, filed his notebook in a drawer of his desk and went on to other things. For nearly 20 years the planets sailed along according to formulas known only to their introverted author.

■■■■■■■■■■■■■■■■■■■■■■■■■■■■■■■■

The Principia

The Mathematical Principles of Natural Philosophy, as its title translates, more often referred to simply as the *Principia,* has been called the most important book ever published, the Bible of science, the greatest intellectual achievement of mankind. . . . You get the idea. These are only slight exaggerations. The book remains a treasure of utmost importance and, except for a few elaborations required by Einstein's theory of relativity, startling accuracy.

The fact that the work lingered in Newton's desk for years before being published really comes as no surprise, for in addition to his intellectual prowess the man was renowned for his complete indifference to moral values, social causes, religious doctrines, political positions, philosophical issues, and even the implications of scientific progress. This disdain for the concerns of his fellow man was perhaps best dramatized during a term he spent as a member of Parliament. His only recorded utterance was a request to have the windows opened.

■■■■■■■■■■■■■■■■■■■■■■■■■■■■■■■■

It wasn't until Edmund Halley (1656–1742), discoverer of the famous comet that now bears his name, came to Newton with a planetary motion problem in 1684 that the law of universal gravitation was finally divulged. In order to straighten out Halley, Newton casually laid before him the cosmic blueprints. Halley was dumbstruck. He insisted that his reticent friend share his sublime secret with the rest of the world, agreeing to assume all editorial duties and the cost of publication himself. Three years later, at the hand of Newton and under the guidance of Halley, *Philosophiae Naturalis Principia Mathematica* became public knowledge.

The *Principia* enjoyed immediate success and provided a solid, long-lasting foundation on which the entire field of classical, or Newtonian, physics still rests. In it the far-reaching concepts of time, space, mass, and motion, as well as their interrelationships, were precisely and mathematically defined for the first time and the law of universal gravitation unveiled. Astronomy was no longer an exercise in mapping the sky, but a true science.

Newton explained that every object in the universe, whether it be an apple or a planet, is attracted to every other object in the universe by means of the mysterious but undeniably real force of gravity. This attraction is a function of two variables, the first of which is mass. The attraction among objects increases in accordance with their collective masses. The earth, an apple, and a watermelon are all drawn to one another by gravity. Because the mass of the watermelon exceeds that of the apple, though, the attraction between the earth and the watermelon is greater than that between the earth and the apple. This results in the watermelon being heavier than the apple. There's also an attraction between the apple and the watermelon, but because of the small amount of mass involved this force is negligible.

The second variable is distance. As gravitational attraction among objects increases with their masses, so it decreases as the distances between them grow. The farther from the earth's center an apple or watermelon is carried the less it weighs. Eventually an object reaches a point where it's attracted as much to the other objects in the universe as it is to planet Earth. At this point it becomes weightless.

The law of universal gravitation resolved once and for all the debate over which was the center of the solar system—the earth or the sun. The sun, with a mass many times greater than that of all the planets combined, was the undisputed pivot point. With its massive gravitational force it whipped the planets around it like a collection of tethered stones. Kepler's ellipses were explained as the inevitable results of the accelerations and decelerations of the planets as they swung around the sun according to Newton's calculus. Minor irregularities of the planetary orbits, unexplained by Kepler, were shown to be result of the gravitational interactions among the planets. The tides were at last understood as the interplay of the gravitational forces of the sun and the moon tugging on the oceans. (See Unit III.14.) Suddenly everything made sense. Everything obeyed the three laws of motion and the law of universal gravitation.

III.9 CELESTIAL MECHANICS

The universe still operates pretty much according to the principles laid down by Newton over 300 years ago, and most likely will continue to do so for the foreseeable future. The laws he penned have been refined and revised somewhat—most notably by the relativity theories of Albert Einstein—but remain, on the whole, a singular accomplishment of human thought and a cogent explanation of celestial mechanics.

Celestial mechanics, like its terrestrial counterpart, operates according to Newton's three laws of motion. (See Unit II.6.) These laws, whether applied in space or on earth, deal with the application of forces to bodies. In the case of terrestrial mechanics these forces can be anything from a kick in the pants to an atomic-bomb blast, and the objects can be an apple, a watermelon, or a Boeing 727. In the case of celestial mechanics, though, the forces to be dealt with are, almost without exception, gravitational ones, and the bodies to which they're applied are either stars, planets, or satellites.

Stars are, for the most part, massive balls of fiery gases that radiate enormous amounts of energy. They tend to be the largest members of the cosmic population. Planets are usually smaller than stars, nonluminous, and associated with a particular star about which they orbit. Satellites are smaller yet, also nonluminous, and associated with a planet the way a planet is associated with a star. The most familiar example of a star, of course, is the sun. It's huge, hot, and surrounded by a cluster of planets. The earth is one of these planets. It's much smaller than the sun, has no radiant energy, and is, in turn, orbited by the moon, which is a satellite. The moon is dark, like the earth, and only a fraction of its size.

The relative positions and motions of the sun, the earth, and the moon were first accurately deduced by the Greeks. It wasn't until Newton reconciled the fall of an apple with the orbit of the moon, though, that their movements were reduced to logical, accurate, and comprehensible terms. Newton discovered that the orbit of the moon about the earth, the earth about the sun, and all other celestial motions as well were the mathematically inevitable results of physical forces acting on moving bodies according to the three laws of motion.

All orbits are, in essence, a near-perfect balance between two opposing phenomena: inertia and gravity. (An exact balance would result in a circular orbit of Aristotelian perfection. The slight inequality of the forces is what results in an ellipse.) Inertia, first described by Galileo, is an example of Newton's first law—an object in motion tends to remain in uniform motion. (See Units II.5–6.) Gravity is an example of Newton's second law—an external force that tends to change the course of an object in motion.

Due to inertia any object tends to maintain a constant velocity along a straight course unless and until acted upon by an unbalanced external force of some kind. A bullet fired from a gun has inertia. In the absence of any interfering forces it will fly forever in the direction it's aimed and at its muzzle velocity. It doesn't do so under normal circumstances because it's subjected to two external forces that, according to Newton's second law, alter its course and velocity. The unbalanced force of friction generated as the bullet pushes its way through the atmosphere persistently and gradually diminishes its speed. The force of gravity alters the bullet's course rather than its speed, pulling it constantly downward.

Ignoring air then, which is only proper when considering celestial motion in the vacuum of space, a bullet fired horizontally from a gun will travel at a constant speed until gravity pulls it to the earth's surface. The force of gravity acts independently of the bullet's forward inertia and will, in fact, pull all bullets downward equally regardless of their forward speed. A bullet dropped from the tip of a rifle barrel may take a second to hit the ground. One fired from that same barrel at 500 kilometers per hour will also take a second to fall. So will one fired at 1,000 kilometers per hour. But, while each of these bullets falls the same vertical distance in the same amount of time, they each cover different amounts of horizontal distance while doing so. The dropped bullet covers no horizontal distance at all, the fired bullets quite a bit, depending upon their velocities.

As bullets are fired at greater and greater speeds, covering greater and greater horizontal distances before being pulled to the ground by gravity, something interesting happens. The curvature of the earth comes into play. Bullets fired with an extreme forward inertia travel over the horizon while being pulled to the earth. They partially circle the globe. These bullets have farther to fall than do slower bullets because their flight paths extend over the horizon and out into space as the surface to which they fall slopes away beneath them. A bullet with enough forward inertia will never actually land. Its forward speed will continually carry it over a perpetually receding horizon as fast as it falls. This bullet will circle the globe, return to its starting point still traveling at the same speed, and will continue to make these round-trips forever. This bullet is in orbit.

The moon circles the earth in a similar fashion. The moon has a certain amount of forward inertia that tends to propel it away from the earth and out into space along a linear path. But the gravitational forces between this satellite and its planet tend to draw the two bodies together. So the moon is forever falling toward the earth while at the same time soaring off into space. The two competing tendencies combine and result in the lunar orbit. (See Figure III.1.) The earth and the sun have a similar relationship, which results in the former's orbit about the latter. In each case the less massive body revolves around the more massive and relatively stationary one. The size of the orbit, the speed of the orbiting body, and the amount of time needed to complete one revolution are all natural consequences of the amounts of inertia and gravity involved.

When gravitational forces exceed inertia the two interacting bodies eventually collide. A bullet with insufficient speed will fall toward and then strike the ground. When inertia exceeds the force of gravity, on the other hand, the two objects eventually part company. A bullet with enough inertia not only to balance gravity but to overcome it—a bullet with escape velocity—will leave earthly environs and fly off into space, never to return.

III.10 THE EARTH IN ORBIT

Our planet, called Earth, is, like all other planets, a more or less spherical object. The spherical shape of heavenly bodies in general is a function of gravity. The individual particles of matter of which a star, a planet, or a satellite is composed are all mutually attracted via universal gravitation. Each is drawn to the other according to its mass and proximity. The net result is that every individual particle is drawn to a central point within the mass known as the center of gravity. Due to the nature of space only one particle can actually occupy this theoretical focal point. The others, like football players converging on a fumble, must content themselves with crowding as close to that point as possible. In the process of converging, the particles assume the format of a sphere, the geometric shape that provides the optimal location to the greatest number of particles.

This great ball of dirt and water, as suspected by Aristarchus, theorized by Copernicus, advocated by Galileo, and explained by Newton, does indeed orbit

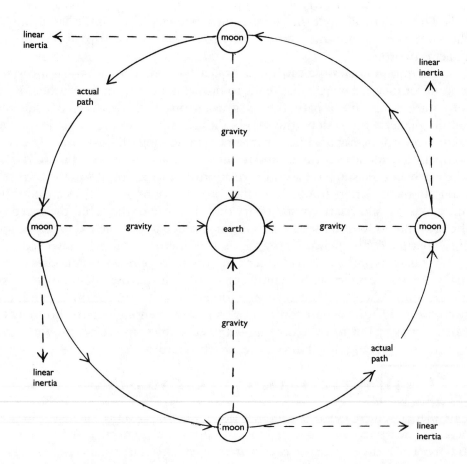

Figure III.1   The mechanics of the lunar orbit

the sun despite its apparent motionlessness. The earth completes one circuit along this more or less circular path every year. In order to cover that much distance in that amount of time the earth must travel at an average velocity of over 100,000 kilometers per hour, or about 300 times the speed of sound.

The fact that there's absolutely no perceptible indication of this headlong rush through space is what made Aristarchus a laughingstock, Copernicus a scared old man, and Galileo a heretic. Their detractors argued that such a speed would leave clouds, birds, lofted missles, and anything else not securely fastened down trailing behind a disappearing earth. Threading a needle, pouring a glass of wine, or even walking down the street, they contended, would be impossible while hurtling through space. But, as first explained by Copernicus, the motion of the earth around the sun in no way affects the operation of terrestrial mechanics. The thread, the needle, the glass, the wine, the street, and the walker are all traveling in unison. Their motion relative to one another is all that matters. Also traveling in unison with the earth is its atmosphere. Within this seamless envelope of air the clouds, birds, and lofted missiles behave without reference to the earth's orbital motion.

So utterly unaffected are events on earth by its revolution around the sun, in fact, that Copernicus, Galileo, and their cohorts were left without any conclusive arguments to support their moving-earth viewpoint. We now have two. The first proof of the earth's motion is called the aberration of starlight, discovered in 1728. According to the nature of electromagnetic radiation, light travels in a straight line. A distant star is visible because its light has traveled directly from it to us. If the earth were stationary, light from a distant star would fall straight down a telescope pointed at it. In fact this isn't the case. To best observe a star a telescope must be held at an ever-so-slight angle to the line of sight. The reason for this is the motion of the earth through space as it orbits the sun.

Consider a pedestrian standing out in a shower of vertically falling raindrops. To optimally protect himself he simply holds his umbrella straight above him, for he experiences the rain as coming from directly overhead. But suppose that this pedestrian sets off on a brisk stroll. Now, in order to compensate for his forward motion through the rain, he must hold his umbrella somewhat in front of him and tilt it a bit in order to best keep himself dry. By moving himself he experiences the rain at somewhat of an angle.

The pedestrian in this illustration is an earthbound astronomer, the umbrella is his telescope, and the raindrops are starlight. If the astronomer and the earth on which he's perched were motionless, he could point his telescope directly into the starlight the way the standing pedestrian aims his umbrella directly at the falling rain. But the astronomer needs to adjust his telescope in much the same way the strolling pedestrian needs to tip his umbrella. The pedestrian needs to cock his umbrella to compensate for his motion relative to the falling rain, and the astronomer needs to angle his telescope in order to compensate for his motion relative to the falling starlight.

●●

Planet Earth

It's now pretty well established that the earth, as suspected ever since the Greeks watched their ships sail out to sea, is round, or at least very nearly so. In fact, the globe is flattened a bit at the poles—sort of like a basketball with someone sitting on it. This irregularity is so minor, though, that it would go totally undetected if the world were reduced to an accurate desktop-sized model. On such a scale Mount Everest and the Grand Canyon would go similarly unnoticed by even the most sensitive fingertip.

In its life-sized version the earth is big enough so that a car constantly cruising at highway speed would need about two and a half weeks to circle it, ignoring rest stops. The fact that 70 percent of the globe is covered by water makes this feat highly unlikely, but even a jet flying at the speed of sound would need almost two days to make a round-trip.

The earth is compressed into the shape of a sphere by the force of gravity. Its densest components tend to sink to its center, or core, while its least dense float on its surface, or crust. The tremendous weight of its outer layers result in extremely high temperatures in its interior.

This great big ball, despite the efforts of the Inquisition, orbits the sun once a year, rotating on its axis once a day as it does so. Its path through space, as Kepler determined, takes the form of an ellipse, which is distorted a bit by the gravitational tug of the moon and neighboring planets. The earth also wobbles somewhat as it spins on its axis. All in all, it's an extremely complicated arrangement, and the accomplishments of those who deciphered it by merely tracking a few specks of distant light are truly remarkable.

—Where will a 1-kilogram bar of lead weigh more, in a valley or on an adjacent mountaintop? How will its weight at the North Pole compare to its weight at the equator?

—What physical events demonstrate that the interior of the earth is hot? What happens when interior strains become greater than the crust can tolerate?

—The earth's distance from the sun varies by only about 3 percent during the course of a year, not enough to have any noticeable effect on our weather. What, then, causes the seasons?

—The nighttime sky undergoes a steady change throughout the year, as constellations drift into and then out of view. Why?

●●

The second proof of earthly motion is called stellar parallax. If you take a stroll around a city park the surrounding high-rise apartment buildings seem to shift positions relative to one another as you do so. The earth in orbit around the sun should similarly provide us with ever-changing perspectives of the distant stars. This fact has been known since the days of Kepler and Galileo, both of whom fervently sought evidence of stellar parallax. Their telescopes, though, weren't capable of detecting any shift in the celestial landscape. It was simply too far away and the earth's orbit too small by comparison. In the year 1823, however, the instrumentation needed to observe, measure, and document minute amounts of stellar parallax was developed. It does exist and confirms the fact that the earth travels about the sun.

III.11 THE ROTATION OF THE EARTH

Common sense has always rather convincingly argued in favor of a stationary earth. It seems like we're standing still. The aberration of starlight and stellar parallax, though, prove that the earth travels through space and that the evidence to the contrary is illusory. Common sense also has historically been in favor of a moving sun. This assumption, too, is as erroneous as it is obvious.

From ancient times onward the predictable appearance of the sun in the east and its disappearance in the west suggested one of two things: either there was a never-ending supply of identical suns that passed overhead one at a time or the same sun made daily excursions. The Greeks, able to prove that the earth was round, decided in favor of the second scenario. The daily alternation of light and darkness, they said, was the result of the sun continuing on its course below the horizon, circling beneath the earth, and then popping up again on the other side. A moving sun made every bit as much sense as a stationary earth and seemed virtually beyond doubt.

But there was doubt. Copernicus doubted. With the given geocentric setup the motions of the planets were simply too complicated for his liking. By holding the sun steady and putting the earth and the rest of the planets in orbit around it he found he could better plot the heavens. This he decided to do. There was a trade-off, though. By swapping a geocentric arrangement for a heliocentric one he simplified the explanation of Mars, Jupiter, Venus, Saturn, and Mercury, but greatly complicated the explanation of day and night.

Copernicus, in order to make day and night possible, had to give to the earth not only a yearly revolution about the sun, but a daily rotation as well. Were it not for this added rotational motion, one side of the planet would experience perpetual day and the other perpetual night. Just as a race car circling a track has one side always facing the infield and the other side always facing the grandstand, so one side of a nonrotating earth would always face the sun and the other side would always face away from it.

According to the scheme devised by Copernicus, then, the sun doesn't orbit the earth from east to west. Rather it's the earth that wheels from west to east as it spins. Like the motion of the earth along its orbit, this rotation isn't consciously felt or normally experienced. You can, with a bit of effort, watch a sunrise and imagine not that the sun is ascending but that you're being carried toward it by a massive, inexorably rolling ball beneath you. But this is a difficult exercise and one that most of Copernicus's contemporaries found impossible, needless, and just plain absurd. They needed some sort of tangible proof.

This was first provided by a Frenchman, Jean Foucault (1819–1868). In 1851 he suspended a pendulum from the dome of the Panthéon in Paris, set it into motion, and observed it for 24 hours. What he saw was that the swing, which he initiated in a north-south orientation, gradually altered its course without any apparent reason. Foucault correctly deduced that there was no reason for the path of the pendulum to change, and that, in fact, it really hadn't. What had changed was the position of the Panthéon, Paris, Europe, and the rest of the planet beneath it. This change of position he properly interpreted as resulting from the rotation of the earth. (See Figure III.2.)

As the earth pirouettes it does so about an imaginary line called its axis. This axis extends from the North Pole to the South Pole and passes through the very center of the planet. It serves as a sort of rotisserie spit upon which the earth spins as it's grilled by the searing heat of the sun. The poles themselves then, along with any internal point directly on the axis, move very little under the influence of the daily rotation but merely turn in place. Other points on the globe's surface trace out circles around the axis every day. The size of the circle circumscribed by any particular point depends upon its distance from either of the poles. The greater the distance from the poles, the larger the circle. The largest circle possible is traced out by points that lie equally distant from the two poles. This circle is called the equator. Any point lying on the equator travels a rotational distance equal to the circumference of the globe every 24 hours.

The poles, then, go virtually nowhere as a result of the earth's rotation while equatorial regions travel somewhere around 40,000 kilometers a day, or 1,700 kilometers per hour. This difference in speeds has a few subtle effects on terrestrial affairs. Things situated near the equator are subjected to a good deal of rotationally supplied centrifugal force. Being whipped along at considerable speed, they have an ever-so-slight tendency to fly off into space. This gives objects a marginally lower weight at the equator than they would have at the poles. Already being whipped along at 1,700 kph, rockets launched from a tropical site need a bit less thrust to obtain orbit or escape velocity than do those launched from polar regions. Even the stuff of which the planet is made responds to this centrifugal force. There's a slight equatorial bulge to the globe and it's the result of tropical regions responding to the outward thrust of rotational inertia.

It's sometimes necessary to compensate for the different rotational speeds

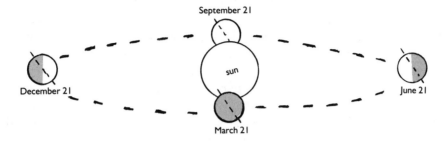

a "side" view of the earth's orbit around the sun

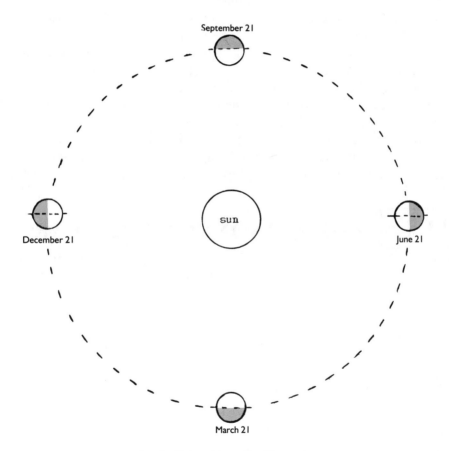

an "overhead" view of the earth's orbit around the sun

Figure III.2   The tilt of the earth's axis

of the different areas of the globe. A missile launched from equatorial Nairobi and intended for Moscow directly to the north, for example, will miss its target if aimed due north. The rotational speed of the African launch site is greater than that of the European target, and the missile will be carried off course to the east as it's launched. In order to strike Moscow the missile must be aimed a bit to the west. North-south airline flights must also take the rotation of the globe beneath them into account when charting their courses.

Of interest to those who are neither airline pilots nor ballistics experts is the Coriolis effect, the influence exerted by the earth's rotation on the weather and other natural events. As a rule, equatorial areas receive more direct sunlight than do polar areas and are therefore warmer. (See Unit III.12.) Air warmed at the equator rises and cooler polar air moves toward the earth's belt to take its place. Like a missile or an airliner, though, this traveling air is carried slightly off course as the world spins beneath it. This results in a deflection of these air masses and a general swirling pattern of air currents. In the Northern Hemisphere this swirling takes a counterclockwise direction and in the Southern Hemisphere it takes a clockwise direction. The swirling is most dramatically illustrated by hurricanes, typhoons, cyclones, and tornadoes.

The Coriolis effect also explains the swirling motion of water as it funnels down the drain of your kitchen sink as well as many prevailing ocean currents. Additionally, it has been proposed as an explanation for a wide variety of natural phenomena including the spiral shape of sea shells, the arrangement of leaves and flower petals, and the migratory instincts of some species of fish and fowl.

III.12 THE SEASONS OF THE YEAR

As evidenced by the aberration of starlight and stellar parallax, the earth hurtles through space as it orbits the sun. As evidenced by Foucault's pendulum and the Coriolis effect, the earth rotates on its axis as it travels along this solar orbit. These are its only two significant motions. The revolution of the earth around the sun, by itself, results in no notable natural events. The rotation of the earth, by itself, results in the daily cycle of darkness and daylight, the swirling patterns of air and water, and a few other minor natural phenomena. What is it, then, that produces the seasons of the year?

The answer to this question lies in the way the earth's revolutionary motion and its rotational motion are related to one another. The plane in which the earth orbits the sun is called the ecliptic. The ecliptic is flat, stable, and contains all points of the orbit as well as the the thing being orbited. It's the imaginary racetrack along which the earth cruises. The imaginary rotisserie spit upon which the earth rotates is called its axis. It neatly impales the earth from north to south and remains fixed in regard to the ecliptic. But—and this is important—the earth's axis is not exactly perpendicular to its ecliptic. If the ecliptic is a horizontal racetrack the axis isn't a vertical flagpole, but a leaning one.

The earth's North Pole is sometimes tilted toward the center of its orbit and sometimes away. When tilted toward the center—toward the sun—the North Pole receives sunlight throughout the course of an earthly rotation. The South Pole at this time is angled away from the sun and receives no sunlight at all during a rotation. Under these conditions the northern latitudes enjoy long hours of sunshine and warm temperatures while regions south of the equator experience diminished supplies of radiant heat and light. This situation corresponds to the Northern Hemisphere's summer and the Southern Hemisphere's winter. The situation is completely reversed six months later after the earth has completed half a circuit around the sun. Then it's the southern extremities that are cocked toward the sun and the northern ones that are directed toward the icy blackness of outer space. Equatorial regions are affected very little by this pitching toward and away from the sun, and as a result have no dramatic changes of season.

As spaceship earth shuttles along its orbit its axis assumes positions that point neither directly toward nor directly away from the sun. Halfway along this route the axis is tilted alongside the sun. At these times most points on the globe receive 12 hours of sunlight and 12 hours of night. At either of the poles the sun is seen to skim along the horizon.

The poles get massive doses of solar heat and light during their summers and hardly any at all during their winters, equatorial regions get a rather steady daily supply throughout the year, and temperate areas have their sunlight proportioned somewhere in between these extremes. But, in terms of total hours of sunlight, all receive more or less equal amounts during the course of a year. It's not the gross amount of sunlight, then, that keeps the poles frigid and the equator tropical.

It's the quality of sunshine that makes the difference. At the equator the daily path of the sun carries it high overhead. Sunlight strikes this part of the globe at a fairly perpendicular angle. At the poles, meanwhile, the sun's path is very low in the sky, even during the height of summer. Rays of sunlight strike these polar areas at very oblique angles, high noon at the North Pole being similar to early morning or late afternoon at more temperate latitudes. These glancing rays are less effective for heating purposes than their more direct counterparts for three reasons.

First and foremost, sunlight is spread out and diluted when it strikes a surface at an angle. Shine a flashlight straight at a wall, then angle it along the wall. The flashlight, the beam, and the amount of energy involved all remain unchanged, but the patch of wall being lit grows from a small, bright circle to a much larger, relatively dim, drawn-out oval. Each illuminated part of the wall receives a smaller portion of the flashlight's output when subjected to angled light than when subjected to direct light. Polar regions similarly receive oblique

● ●

Orbital Landmarks

As the earth makes its yearly trip around the sun it passes four special reference points along the way. These are the summer and winter solstices and the vernal and autumnal equinoxes.

In the Northern Hemisphere the summer solstice occurs on or about June 22, the autumnal equinox on or about September 23, the winter solstice on or about December 22, and the vernal equinox on or about March 21. These are customarily called the first days of summer, fall, winter, and spring, respectively. In the Southern Hemisphere the dates are the same but the labels reversed.

On June 22 or so the Northern Hemisphere receives its maximum quota of daily sunlight and sees the sun as high in the sky as it will ever get. All points lying north of the Arctic Circle experience 24 hours of daylight on this date while those south of the Antarctic Circle experience 24 hours of darkness. Also on this date, all points lying between the equator and the Tropic of Cancer experience at least a moment of vertical sunlight. Six months later it's the Southern Hemisphere that basks in sunlight while the Arctic remains dark.

The tilt of the earth's axis at an angle of 23°27' to the ecliptic is the sole cause of these seasonal variations. The degree of tilt is what results in the Tropic of Cancer being drawn at 23°27' north of the equator and the Tropic of Capricorn at 23°27' south latitude. The Arctic Circle, meanwhile, lies at 66°33' north latitude, or 23°27' south of the North Pole, and the Antarctic Circle is similarly situated in regard to the South Pole.

These four circles divide the globe into five climatic regions. Everything north of the Arctic Circle is called the North Frigid Zone, everything south of it to the Tropic of Cancer is called the North Temperate Zone. These are matched by the South Frigid and Temperate Zones in the Southern Hemisphere. Between the Tropics of Cancer and Capricorn lies the Torrid Zone.

—Why do the dates of the solstices and equinoxes vary a bit from year to year?

—Is it possible to use a sundial to tell time in the Frigid Zones? What about at the equator?

—Why are solar panels generally tilted instead of laid out flat? Which way do they point in the Northern Hemisphere?

—What are the effects of Daylight Savings Time on the distribution of sunlight?

● ●

rather than direct sunlight and are warmed to a lesser degree than those areas receiving overhead sunlight.

Secondly, oblique light travels a greater distance through the atmosphere than does overhead light. Perpendicular light takes the shortest available path through the air, like a pedestrian crossing a street at a crosswalk. Oblique light, in contrast, takes a longer route, like a pedestrian crossing a street diagonally. During its longer time spent traversing the atmosphere oblique light loses more of its energy to the air, where it's quickly dissipated, than does overhead light. Less radiant heat reaches the earth's surface and less warming takes place.

Thirdly, while the angled light passing through the atmosphere is partially absorbed, other rays of oblique light are totally deflected by the atmosphere. They bounce off the upper layers of the atmosphere like stones skipping off the surface of a still pond and are forever lost to the emptiness of space. More vertical rays, like dropped stones, aren't deflected in this manner.

Without the tilt of the earth's axis December wouldn't be noticeably different from June, our yearly trip around the sun would be little more than an academic exercise, and our whole sense of time would undoubtedly be altered. The equatorial zones would still be hot, the polar ones cold, and those in-between temperate, but there would be no rhythmic ebbing and flowing of long days and short ones, nor of hot seasons and cold ones.

III.13 | THE MOON

Next to the dramatic alternation of day and night caused by the earth's rotation on its axis and the yearly cycle of the seasons caused by the tipping of the earth's axis in relationship to the ecliptic, the most important and obvious astronomical occurrence is the motion of the moon about the earth. (See Figure III.3.)

With a brightness exceeded only by that of the sun, the moon is, and always has been, something special. It's been honored by temples and pyramids; worshipped as a provider of rain, fertility, and virility; and linked, with varying degrees of success, to meteor showers, the tides, and the menstrual cycle. With an ever-changing shape and a variable time of appearance, the moon has also been attributed with a great mystique. It's been feared as a cause of disease, misfortune, and a form of mental instability named lunacy in its honor; named as the source of the supernatural powers of witches and werewolves; and associated, again with varying degrees of success, to suicides, epileptic seizures, crime waves, and sleepwalking.

While some of these lunar functions would appear a bit fanciful, or highly debatable at best, none can be totally dismissed, for the moon, according to Newton's law of universal gravitation, exerts a very real force upon planet Earth. Due to the moon's relative proximity and its large size, this force is fairly powerful as well.

In terms of simple dimensions, the moon is fully one-fourth the size of the earth. Because the moon is far less dense, though, its mass is only 1/80th of the

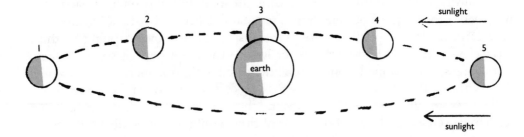

the orbit of the moon around the earth

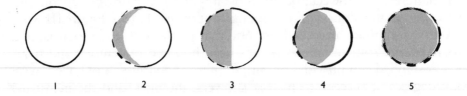

the moon as seen from the earth

Figure III.3   The phases of the moon

earth's. This is still a considerable amount, however, and exceptional for a satellite. The moon is, in fact, the largest known satellite, larger than several bona fide planets and, in terms of size compared to affiliated planets, in a class by itself. Other planets may have multiple moons, prettier moons, or even rings, but none has an orbiting dependent that so nearly approaches its own size and mass. It's even been suggested by some astronomers that the moon not be classified as a true satellite of the earth, but that the earth and moon together be thought of as a sort of double planet. Although not likely to catch on, this idea has some merit, for the earth and moon orbit the sun in tandem, their paths intertwined, swinging about a mutual center of gravity like a giant, lopsided dumbbell.

The size of the moon is especially meaningful within the context of its location. A body the size of the moon would have a negligible gravitational effect on the earth were it as distant as the sun, say, or another planet. But the moon, in astronomical terms, is virtually on top of us.

It's the moon's nearness that accounts for its brightness. The moon, being a satellite, is a nonluminous object. Like a massive boulder floating in space, it emits no light of its own, but is illuminated by the light of the sun and is thus visible. Moonlight, in fact, is just sunlight reflected off the moon to the earth. As Galileo discovered, the moon's surface is rugged, raw, and rock-strewn— not the most desirable qualities for a reflector. Still, because it fills so much of the sky, especially when full, the moon reflects enough light to brighten an otherwise black night. Even a less than full moon provides more light than all the stars combined.

The moon, despite its constantly changing appearance, always presents the same side toward earth. The new moon, first quarter, full moon, third quarter, and all the other intermediate phases are just different portions of a single face. The moon, like the earth, spins on an axis. The earth spins on its axis once a day while it orbits the sun once a year. The earth has 365 axial rotations per solar orbit and can be seen to spin when viewed from the sun. The moon, though, takes exactly as long to spin on its axis as it does to circle the earth. It always keeps its familiar face directed squarely at the center of its orbit. When viewed from the earth it appears motionless on its axis. This interlocking of the moon's period of rotation and period of revolution is not just a coincidence, but a result of the earth's gravitational force. This pull is so strong that it prevents any further spinning of the moon relative to the earth.

We on earth, then, are limited to a view of only one side of the moon. The other side is always turned away from us. Although often referred to as the "dark" side of the moon, this hidden portion of the moon is more correctly called the far side of the moon for it receives just as much sunlight as does the familiar, or near, side. As is the case for the earth or any other body near any other luminous object, half the moon is always exposed to the sun and half shielded from it. And, because the moon spins on its axis, the half exposed to the sun is continually changing.

The moon's cycle of light and darkness, like that of the earth, is equal to the time it takes to spin around once on its axis. For the moon the period of rotation on its axis, like its period of revolution around the earth, is approximately one month—a term derived from the word *moon*. About once a month—during a full moon—we on earth see the entire illuminated half of the moon. We see the entire expanse of lunar day. At other times we see part of the illuminated half of the moon and part of the shaded half. And sometimes— during a new moon—we see only the unilluminated portion of the moon. The side we see then is being subjected to lunar night while the far side is flooded with sunlight.

III.14 THE LUNAR ORBIT

As does the earth's orbit about the sun, the moon's orbit about the earth closely approximates a circle. The moon's plane of ecliptic about the earth also closely approximates the earth's plane of ecliptic about the sun. If a line connecting the sun and the earth is labeled horizontal, then a line connecting the earth and the moon is also nearly horizontal. The moon also circles the earth in the same direction as the earth spins on its axis and circles the sun. Taking the North Pole as up, an overhead view of the entire system would reveal the earth circling the sun in a counterclockwise direction, the moon circling the earth in a counterclockwise direction, and both bodies turning on their axes in counterclockwise directions as well.

The lunar orbit takes a little over 29 days. During each day, then, the moon completes just about 1/29th of its orbit around the earth. As we on planet Earth spin on our axis we temporarily lose sight of the moon, then see it again the next day. Because the moon has moved along its orbital path during this time we have to complete slightly more than one full rotation in order to regain the perspective we had the previous day. In fact, we have to complete about an extra 1/29th of a rotation in order to compensate for the 1/29th of a circuit about us the moon has made. This extra 1/29th of a rotation takes 1/29th of a day, or approximately 50 minutes. Every day we catch sight of the moon 50 minutes later than we did the day before. Every day moonrise occurs about 50 minutes later than it did yesterday and about 50 minutes earlier than it will tomorrow.

The changing shape of the moon is also due to its monthly circuit of the earth. At the time of the full moon the sun, the earth, and the moon lie more or less along a straight line with the earth in the middle. As we look at the full moon the sun is at our backs and the side of the moon facing us is also facing the sun and thus fully illuminated. The sun and the moon at this point in the lunar cycle are opposite one another with regard to the earth. As we rotate toward the east we catch sight of the moon just as we lose sight of the sun. Because of this arrangement the rise of the full moon always takes place right around sunset.

■■■■■■■■■■■■■■■■■■■■■■■■■■■■■■■■■■

Moonlight

The rise of the full moon and the setting of the sun coincide most closely at the time of the equinoxes. It's then that moonrise is especially spectacular. The full moon nearest the Northern Hemisphere's autumnal equinox—September 23 or so—has won particular fame as the "harvest" moon. But neither this nor any other full moon is quite as bright or as big as might be supposed.

Even at its fullest, the moon pales in comparison with the sun. It would, in fact, take something on the order of a half-million full moons—a whole skyful of them—to equal the brightness of the sun. It's only the surrounding blackness that makes the moon seem brilliant.

The size of the moon as it rises is also an optical illusion. Careful measurements have determined that the moon is no larger when on the horizon than when higher in the sky. It's just that when it's lower in the sky it's subconsciously associated with the much closer terrestrial objects alongside of which it appears. Its distance is thus misjudged and its size miscalculated by the eye.

■■■■■■■■■■■■■■■■■■■■■■■■■■■■■■■■■■

The day after a full moon, moonrise occurs some 50 minutes after sunset. By the time the moon comes into view the sun has already been sinking below the horizon for almost an hour. The sun and the moon are no longer directly opposite one another. The side of the moon facing the earth and the side facing the sun are no longer identical. The moon appears slightly less than full because it's being lit not from directly behind us, but from behind and a bit off to one side. A small portion of what faces us is unlit and invisible. A small portion of what is lit lies on the far side of the moon.

The retardation of moonrise and the shrinkage due to sidelighting—the waning of the moon—continue up until the time of the new moon. Then the earth, the moon, and the sun are once again directly aligned, but the moon, having completed half an orbit, has now moved from the outside position to a point between the earth and the sun. From earth, the sun and the moon appear more or less on top of one another and rise and set in unison. The moon now is invisible for two reasons. In the first place, it occupies the sky at the same time as does the sun and, like the stars, is rendered insignificant by its brilliant companion. In the second place, the moon is now completely back-lit. The side facing the earth is turned away from the sun and unilluminated. Like the full moon, though, the new moon lasts briefly. Soon the moon is rising and setting later than the sun and receiving a small degree of sidelighting as it waxes once again.

Sometimes, during a full moon, the alignment of the sun, earth, and moon is so precise that the shadow of the earth falls across the moon. The moon receives no sunlight because it's blocked by the intervening earth. It's darkened during what's called a lunar eclipse. Similarly, during a new moon, the alignment of the sun, moon, and earth may be so exact that the moon moves in front of the sun and blocks it out. It's then the sun that's temporarily invisible during what's called a solar eclipse. Solar eclipses are more common than lunar ones in terms of sheer numbers but, because they affect only a tiny portion of the globe and last only a few minutes, are rarely experienced. Lunar eclipses, on the other hand, can last for as long as two hours and play to a worldwide audience.

Both types of eclipses occur at predictable times according to a recurring schedule, one complete cycle of which takes just over 18 years to complete. The cycle, called the saros, is a result of the complex motion of the moon about the earth and the earth about the sun. Due to complications arising from the irregularities in the shape of the earth and anomalies in its orbit resulting from the occasional presence or absence of other nearby planets, the exact repetition of any given earth-sun-moon configuration requires 18 years and 11 days to reappear. While the saros was undoubtedly decoded by several ancient cultures, it was Thales of Miletus who not only predicted eclipses but explained them in terms of the relative positions of the sun, the earth, and the moon.

It was another Greek thinker, Pytheas (4th century B.C.), who first associated the moon with the tides. Pytheas noticed that the tides, like the appearance of the moon, happened about 50 minutes later every day and postulated a connection between the two seemingly unrelated events. Pytheas, of course, didn't have a clue as to why his theory should work, but work it did. And because it worked it enjoyed a prolonged if sporadic currency among those who were interested in such things.

It wasn't until Newton quietly penned his description of gravity that a logical explanation could be offered, however. According to his universal gravitation the moon attracts the earth the same way the earth attracts the moon. The waters of the oceans, being liquid, are more able to respond to this lunar tug than are the solid continents. They're more drawn toward the moon than is the planet as a whole, and this is what causes the tides.

There are, in most places, two tides every day. More accurately, there are two tides every 24 hours and 50 minutes, equally spaced every 12 hours and 25 minutes. Each day the cycle of these tides lags another 50 minutes behind the clock, just as does moonrise. The two tides can be thought of as two standing waves, one pointed more or less at the moon overhead, the other, on the other side of the globe, pointed more or less away from it. As these tidal waves follow the moon in its 29-day orbit the earth rotates beneath them once a day, turning from west to east. The tides are then experienced as sweeping across the globe's face from east to west. The wave directed toward the moon is the result of the moon's somewhat greater attraction for the relatively nearer oceans than for the relatively farther away earth beneath them. The wave pointed away from the moon is the result of the moon's somewhat greater attraction for the earth in general than for the marginally more distant oceans on its far side.

As any coast dweller knows, the size of the tides varies greatly even as their schedule remains quite regular. This is the result of the gravitational force of the sun either reinforcing or partially canceling that of the moon. During full and new moons, when the earth, moon, and sun are aligned, the sun's tidal pull is exerted in the same general direction as is the moon's. These conditions produce very extreme, or spring, tides. During the moon's first and last quarters the sun and the moon are located at right angles to one another with respect to the earth. At these times the sun's tidal influence somewhat offsets that of the moon to produce very gentle, or neap, tides. The tides are further influenced by other planets, especially Venus, which is very close, and Jupiter, which is very large. Tides are also affected by weather, local geography, and a number of other factors, some of which are understood and some of which aren't. Their association with the moon, though, is no longer a far-fetched theory but a well-established fact.

III.15 THE SOLAR SYSTEM

Although far and away the most important heavenly bodies to us earthlings, the sun and the moon are by no means the only ones about which we should be concerned or curious. Jupiter and Venus, due to their size and proximity, have enough influence on our planet to notably affect the tides. They may have other as yet undiscovered but by no means trivial impacts as well. The other planets, too, may hold some sway over our condition. Newton's law of universal gravitation, remember, states that every body in the universe, no matter how small or how distant, influences every other body in the universe through the force of gravity.

The entire cosmos, according to Newton's theory, is like a single, enormous, intricate, three-dimensional spider's web. Strands of gravitational attraction, some short and strong, others long and weak, connect everything to everything else. Every object of the heavens is suspended in place by a complicated, delicate balance of competing lines of force woven across empty space. Pluck any strand and the whole web responds. To be fair, Newton's law also states that the influence of one object over another diminishes rapidly as the distance between them increases. The objects expected to have the greatest impact on the earth, then, would be those that are closest to it.

Aside from the moon and sun, the earth's nearest neighbors are its fellow planets. The planets, together with the sun they orbit, their various satellites, the asteroids, and an assortment of comets, compose what is known as the solar system. As its name implies, the solar system is based on the sun. The sun is its focal point. At latest count, nine planets orbit around the sun. In order of their average distance from the sun, starting with the nearest, the planets are:

Mercury, Venus, Earth, Mars, Jupiter, Saturn, Uranus, Neptune, and Pluto. Among them these planets claim a total of 31 satellites. Between Mars and Jupiter there lies not another planet, as certain mathematical models suggest there should be, but a band of floating rubble. Called the asteroids or planetoids, these chunks or orbiting debris may be the remains of a one-time planet that disintegrated or the basic ingredients of a would-be planet that failed to coalesce. Also orbiting the sun, but not true planets, are a number of comets. Comets are much, much smaller than planets, have no satellites, and travel in great extended ellipses rather than in approximate circles as do true planets.

Not everything in the solar system falls neatly into one of the above categories. We've already seen how the moon, the largest satellite in the solar system, is really something of a planet-satellite hybrid. Likewise Pluto, the smallest planet, is something of a comet-planet hybrid. Its size and the shape of its orbit both lie somewhere in between true planet and true comet norms. By the same token Jupiter, the largest planet, can be called a star-planet hybrid. Jupiter is nearly as large as some stars, generates a bit of its own heat, and with its collection of 12 attendant moons can easily be thought of as an entire quasi–solar system. The inspiration it gave Galileo was completely justified.

Despite a sizable and varied population of inhabitants, the solar system, like the atom, is mostly empty space. Were it reduced to a scale model 10 kilometers (six miles) across, its largest member, the sun, would be only about the size of a beach ball. Planet Earth would be about the size of a pea on this scale, and would lie about a city block distant from the beach ball. Pluto, at the edge of the mock-up, would be smaller than a grain of rice. In its life-sized version the solar system is almost inconceivably huge. A ray of light leaving the sun requires over eight minutes to reach the earth and nearly five hours to reach Pluto.

The solar system as a whole is shaped something like a saucer. All the planets as well as all their satellites share a similar plane of ecliptic. What's more, they all travel in the same general direction and all their moons in turn orbit them in this same, call it counterclockwise, motion. Nearly all rotation of planets and moons is also done in a more or less counterclockwise manner. The prevalent swirling motion in the solar system lends considerable support to the argument that all its members have a common origin of some kind. (See Unit IV.5.)

In accordance with the operation of gravity, each planet wheels around the sun at its own pace. This pace is a function of the size of the planet's orbit. The planets nearer the sun have a shorter period of revolution, or year, than do those farther away. So the earth is continually being overtaken by its speedier inside neighbors—Mercury and Venus—and is continually overtaking its slower outside neighbors—Mars, Jupiter, Saturn, and the rest. When the inner planets have recently overtaken the earth or when the outer planets are about to be overtaken, they appear in the morning sky and are called—somewhat erroneously because they're planets—morning stars. When the inside planets are about to overtake the earth or when the outer planets have just been overtaken, they appear in the evening sky and are called evening stars. It's this constant passing and being passed, not roaming spirits or epicycles, that accounts for the elaborate paths of the planets across the nighttime sky.

III.16 THE FIXED STARS

For all its immensity, the solar system is just our own front porch when it comes to depicting the vastness of the universe as a whole. If the distances among the members of our little cluster of planets are awesome, the distances to the next nearest known object beyond them are nothing less than astounding. Even when stated in light-years—the distance a ray of light will travel in one year—the numbers get cumbersome to the point of uselessness. It's the magnitude of their distances that makes the so-called fixed stars appear stationary. No amount of motion on our part is capable of changing our perspective on this celestial canopy to any appreciable extent. Even the loop of the earth around the sun is about as effective for this task as craning one's neck to get a better view of the far side of the moon.

The fixed stars aren't really fixed, of course. They just seem that way. Like everything else, they're in states of perpetual motion. These motions, though, are so slight when compared to their distances that they're completely imperceptible to the naked eye. Even a Tycho Brahe, devoting a lifetime to the systematic scrutiny of these tiny points of light, wouldn't detect any noticeable rearrangement among them during the course of his studies. The most mobile fixed star, it's been estimated, will require 18,000 years just to traverse a portion of the sky equal to the size of the full moon. The sky we see, then, is very much similar to the one seen by Ptolemy and Copernicus.

That sky, under optimal viewing conditions, consists of some five thousand or so pinpoints of light visible to the naked eye. They appear in various degrees of brightness and in various hues depending upon their proximity, size, and temperature. Most are single stars something like the sun—giant balls of incandescent gases undergoing a continuous nuclear explosion. Others may be paired up as double or binary stars—two stars swinging around one another like a pair of celestial ice skaters. The most distant, fuzziest specks of light aren't stars in ones or twos, but whole galaxies—huge clouds consisting of millions upon millions of individual stars reduced to a single dot of light by virtue of their staggering remoteness.

Our sun is a member of one such galactic cloud. We've named it the Milky Way. All the visible single stars around us are fellow members of this cloud. The Milky Way we reside in, like an enormous solar system, is somewhat disk-shaped. We occupy a position fairly close to the edge of this disk, embedded in a region fairly sparsely populated with stars. When we peer out into the depths of space we're looking out through this surrounding cloud of stars in various directions. For the most part these views provide fairly similar densities of stars. But when we look edgewise across the disk, toward its center, we see a faint band thickly strewn with stars. This is the bulk of our own Milky Way galaxy.

Any particular view of the nighttime sky depends upon the orientation of the earth and the position of the observer on its face. (See Figure III.4.) Views from the poles are fairly predictable because the earth's axis always points in the same direction relative to the fixed stars. To an observer at the North Pole, for example, "up" is always the same absolute direction because she merely spins in place while the earth rotates inconsequentially beneath her. The North Star, Polaris, will always be directly overhead and surrounded by the same familiar neighbors. As the observer spins with the globe these neighboring stars appear to trace out concentric circles around Polaris. Its nearest neighbors trace out the smallest circles while those stars barely visible at the edge of the sky skim along the horizon. This view remains pretty much unchanged throughout the year except for the periodic entrance and exit of the star we orbit, the sun. A similar situation exists at the South Pole except, because Antarctic "up" is the diametric opposite of Arctic "up," the sky is totally different. Polaris, being straight "down," is never visible from the South Pole.

At the equator, however, the stars, like the sun, rise in the east and set in the west. This is because an equatorial observer doesn't rotate in place, but is swept around the full circumference of the planet from west to east every 24 hours. Instead of "up" being a constant direction relative to the fixed stars it's a constantly changing one. If a North Pole observer looks "up" and a South Pole observer "down," then a tropical observer looks always "out" at a changing panorama. He or she is constantly exposed to new vistas as old ones glide by.

In the great stretch of area between the poles and the equator the situation is a hybrid of the two extremes. From these intermediate stations some stars—those nearest Polaris in the north, those nearest the southern celestial pole, roughly approximated by Crux, the Southern Cross, in the south—can be seen all night and appear to move in complete circles. Stars near the horizon, meanwhile, rise and set as only a portion of their circle is visible. The difference in which stars circle and which rise and set is a function of an observer's nearness to a pole. The change in the Great Bear from being a group of circling stars to stars that set, remember, is what tipped off Thales to the fact that the world was round.

While the circling stars remain pretty much the same year-round, the rising and setting stars don't. As the earth makes its way around the sun the direction of "out" is always changing. Each time night falls the earth has moved 1/365th of the way along its orbit. The "outward" vista has shifted by 1/365th of a circle—more or less 1 degree. A small slice of sky unseen the night before has come into view while another small slice has been lost. As the year passes a changing panorama of stars slowly parades through the portion of the sky most directly "out" from the sun. This panorama repeats itself every year in a predictable sequence of vistas.

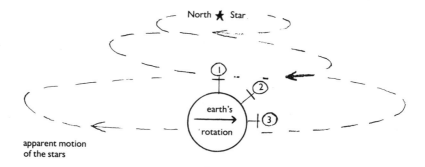

the apparent motion of the stars as a result of the earth's rotation

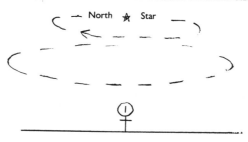

the apparent motion of the stars from the North Pole, position 1

the apparent motion of the stars from the middle latitudes, position 2

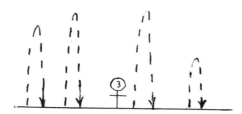

the apparent motion of the stars from the equator, position 3

Figure III.4   Various perspectives on the nighttime sky

■■■■■■■■■■■■■■■■■■■■■■■■■■■■■■■■

The Zodiac

Although of importance to relatively few of us these days, the zodiac has historically been considered a powerful force. Astrologers throughout the ages have believed that, just as the sun brings day and night and the moon rules the tides, the stars, too, have a strong influence on events on earth.

Any modern astrologer should know that the current alignment between the seasons and the zodiac is not what it used to be. This is due to what is called the precession of the equinoxes—the fact that the time between one spring equinox and the next is about 20 minutes short of the time it takes the earth to complete a full circuit of the sun relative to the stars.

This slight irregularity between the calendar, based on the equinoxes, and the zodiac, based on the stars, has been accumulating since the current zodiac was organized some 2,000 years ago, and now amounts to just about one month. This means that the stars that appeared to the Greeks in April and inspired Aries the Ram now appear in May, which is the time of Taurus the Bull.

■■■■■■■■■■■■■■■■■■■■■■■■■■■■■■■■

The Greeks, borrowing liberally from the Babylonians, recognized that each portion of the panorama corresponded to the time of the year during which it occupied center stage. They broke the panorama down into 12 sections, assigned the stars of each section to identifiable groups called constellations, and gave each section a name. The name assigned to each section was inspired by its dominant constellation: Taurus the Bull, Aries the Ram, Leo the Lion, and so forth. This scheme has survived to the present day and is still known by the name the Greeks gave it: *zodiac,* meaning band of animals.

III.17 | THE UNIVERSE

On a clear night, under proper conditions, an earthbound observer can see something on the order of five thousand distant objects sparkling overhead. These five thousand are the tip of an iceberg of truly astonomical proportions. With a decent pair of binoculars the number of visible stars swells to about fifty thousand. With a small telescope the number is again multiplied tenfold. From a sophisticated observatory literally billions of stars can be seen.

Amazingly, these remote objects can not only be seen, but measured, clocked, charted, analyzed, and even weighed. The results yield a universe whose inhabitants increase not only in number, but in variety as well. Besides galaxies, stars, planets, and satellites there are also novas, supernovas, quasars, pulsars, planetary and diffuse nebulae, pulsating Cepheids, red giants, white dwarfs, black holes, and clouds of interstellar dust. They represent, for the most part, different stages in the life cycle of a star as it builds itself up from tiny particles by the action of gravity, burns brightly via nuclear fusion, and finally ends up as a giant cinder of waste products.

Most of what we know about our distant neighbors is based on nothing more than the feeble rays of light that find their way here across the endless expanses of space. Not only is this light strictly limited in quantity, but its quality is also less than it could be. The earth, it turns out, isn't exactly the best place from which to gather starlight. The problem is the atmosphere. The same collection of nitrogen, oxygen, and airborne impurities that blues the sky and yellows the sun also discolors the stars. The same thermally created layers of air that cause mirages also cause the stars to twinkle and the moon and planets to blur. Still, we've succeeded in amassing an amazing amount of information about our far-flung fellows.

By the use of spectra analysis, the elements of which the stars are composed can be determined. (See Unit I.8.) The chemical composition of a star, along with its overall color, can in turn be used to approximate its temperature. Its temperature, coupled with its brightness, is used to estimate its distance. Distance, together with temperature and brightness, helps determine a star's size. Mass is inferred from the strength of a star's gravitational impact on other heavenly bodies. Size and mass together determine density, which offers clues to structure.

Additionally, light, like sound, is subject to the Doppler effect. (See Unit II.19.) Just as an approaching fire truck's siren crowds its emitted sound waves to produce a higher-pitched wail while a receding one strings them out to produce a lower-pitched wail, an approaching or receding star crowds or strings out its emitted light waves. An approaching star produces a slightly bluish light while a receding one appears slightly reddish. The amount of reddening or bluing also indicates a star's speed. And, by comparing the light from opposite edges of a single star, a difference in their rates of approach or retreat can sometimes be detected. Analysis of this difference results in knowledge about the speed, direction, and axis of a star's rotation.

Based on the evidence so far gathered, it appears that the universe as a whole is expanding. Like dots drawn on a balloon being inflated, every object in the universe seems to be moving farther away from every other object. Three theories have been developed to accommodate the general nature of the universe based on this growth. The first, and most popular, is the so-called Big Bang theory. According to this proposal the entire universe originated when a primal cosmic egg hatched to the accompaniment of a huge explosion. The universe, say the Big Bangers, has been expanding like a great shock wave ever since, and will continue to do so indefinitely.

The second theory describes what is referred to as an oscillating universe. The oscillating universe was created in the same way as the Big Bang universe. The difference is that the oscillating universe won't go on expanding forever,

● ●

Three Windows

The earth's atmosphere effectively blocks all but three types of electro-magnetic radiation. It's only through the three "windows" provided by radio waves, infrared radiation, and visible light that we are able to detect objects and events in outer space.

While infrared and radio waves originating in space are of some value, it's light that provides us with the overwhelming majority of our information and, just as it has been ever since Galileo's day, the telescope is the preferred method of observing it.

The critical factor in any visual task is the amount of light available for study. Telescopes are useful in that they act like funnels, gathering light through a wide opening at one end and then concentrating it at a small focal point at the other. The larger the opening a telescope has the more light it can gather and the clearer the image it can produce. The largest telescopes, at the limit of technological and economic feasibility, are capable of gathering about a million times as much light as the unaided human eye.

This advantage is further multiplied by the use of photography. A camera, besides providing a permanent record for later analysis, allows additional light to be collected. Long exposure times result in images being reinforced and clarified as one ray of light is piled atop another. The use of telescopes to make time-exposure photographs necessitates their being mounted on elaborate mechanical platforms. These plat-forms must delicately and precisely twist and turn in order to keep the telescope steadfastly pointed at its target while the earth beneath it rotates on its axis and revolves around the sun.

—Observatories are almost always built on remote mountaintops. Why? What will be an even better location for telescopes of the future?

—The main function of the U.S. Naval Observatory near Flagstaff, Arizona, is to establish correct time. How does it do this?

—Although radio waves provide an idea of how big a body is and where it is, they fail to produce an image. What are the advantages of radio astronomy?

—It seems extremely fortunate that the most useful of all waves, those of visible light, are among the three types to penetrate the atmosphere. Is this a matter of luck? Could it be otherwise?

● ●

but will eventually cease growing and then begin to contract. The contraction, nothing more than the expansion in reverse, will continue until everything is once again reduced to the cosmic egg from which it came and from which it will once again explode.

The last theory is that of the steady-state universe. This universe is undergoing constant expansion, as are the other two, but did not come from and will not revert to a primal source. The steady-state universe has always existed as it is and will continue to do so. To provide for stability in the face of perpetual expansion the steady-state universe constantly generates new matter to keep its growing space populated at a constant density.

Briefly then, the Big Bang universe is a universe with a beginning but no end, the oscillating universe is a universe with a series of beginnings and endings, and the steady-state universe is a universe with neither a beginning nor an end. This sort of speculation may seem like a waste of time to some, and for them it probably is. But to others the contemplation of questions like How big is the universe? Does it have an edge? If so, what lies beyond that edge? and, Is there someone or something out there wondering what I'm wondering? is time spent in the best possible way.

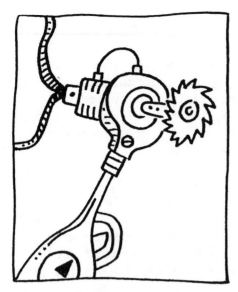

HOW LONG YOU'RE IN THE DENTIST'S CHAIR IS DIRECTLY PROPORTIONATE TO THE TIME IT TOOK YOU TO EAT THE BOX OF SUGAR GUMMY SNAPS THAT PUT YOU THERE

CHEF LOU PROVES THAT GALILEO WAS ABSOLUTELY CORRECT: THE EGGS FALL AT THE SAME RATE AS THE ONIONS AND CHEESE

CLOCK FOR THE REAL WORLD

IV
TIME

"Time wastes things away, and all things grow old through time."
—Aristotle, 384–322 B.C.

IV.1 | THE ESSENCE OF TIME

The astronomical events involving the earth, the moon, the sun, the solar system, the Milky Way, and the enormous stretches of space in between and beyond them all involve motion. The earth rotates on its axis once a day. The moon revolves around the earth about once a month. The earth, in turn, revolves around the sun once a year. The rest of the planets in the solar system also revolve around the sun, each according to its own specific year. The sun, meanwhile, spirals along the edge of the Milky Way and even the Milky Way itself cartwheels through space as the universe as a whole expands. Everything, everywhere is in a constant state of flux.

Terrestrial events as well consist of perpetual motion. Identifiable chunks of matter move when acted upon by unbalanced forces according to the laws of mechanics. Oceans of air, water, or other elastic media ebb and flow in waves as they conduct sound from source to receiver. Heat is the net result of molecular vibrations operating on a massive collective scale. Electricity is the migration of subatomic electrons from one place to another. Light and its related forms of electromagnetic radiation exist as the propagation of energy across expanses of space. And even apparently inert matter is, in reality, a frenzy of atoms each of which is an intricate assembly of swarming components.

Everything, from the tiniest atomic fragment to the most mammoth galaxy, is in motion. And motion, by its very definition—the change of position from one moment to the next of a physical body—involves change. Within this single idea of motion, or change, are embodied all the fundamental elements of physical reality. To begin with, it involves matter. Matter is the stuff that changes position or is altered. Secondly, it involves energy. Energy acting on matter is what causes motion or change to take place. Thirdly, it involves space. It's space that accommodates the various positions assumed by matter during motion. Finally, it involves time. Change is the alteration of matter from one moment to the next, and the existence of these various moments is the essence of time.

Of the four basic constituents of physical reality—matter, energy, space, and time—time is the most intimately related to change. Unlike the other three, time can't even be conceived of in the absence of change. Matter, at a temperature of absolute zero, becomes totally static. Energy, in its potential phase, can remain poised motionlessly. And unoccupied space exists as an inanimate vacuum. But time, in the absence of change, can neither be defined nor exist. A second, a day, a year, or an eternity would be equally meaningless without the passage of some physical event by which to identify or measure them.

Time, then, can be defined as that which is measured by change. But this definition, while technically correct, isn't very satisfying. Time is too elusive to be so neatly disposed of. Matter can be touched, tasted, picked up, put down, stored on a shelf, weighed, recycled, bought, and sold. Energy can be seen, heard, felt, manufactured, transformed, harnessed, and put to work. Space can be occupied, vacated, rented out, subdivided, closed in, or reserved. Each can be somehow directly experienced. Each has a permanence about it. And each has, to a certain extent at least, fallen under our control. Of the four basic elements of reality only time has escaped our meddling. This by itself testifies to the fact that, while we can define time, we don't understand it.

IV.2 | OLDER, NEVER YOUNGER

What it is that makes time so different from the other aspects of reality—matter, energy, and space? The answer to this question must surely lie in the sensation that time is impermanent. Matter and energy, although they undergo radical transformation, can neither be created nor destroyed according to the laws of their conversation. (See Units I.5 and II.2.) And space, according to the theory of the expanding universe, is actually a commodity whose supply is steadily increasing. (See Unit III.17.) Time alone is something that seems to vanish without a trace, something that we can't hoard to use later as we see fit, something we can't redistribute or improve upon, and something that we fear we might eventually run out of.

It is, however, only our access to time that's really restricted. Just as our confinement on planet Earth condemns us to experience only a tiny volume of space, so our existence as mortal beings condemns us to experience only a tiny period of time. But time itself, like space, is in no way limited. No matter when or where changes occur they'll take place within the context of time. Time in this sense is more properly called time interval. Time interval stretches perpetually throughout the duration of all change. Time interval, as far as we know, is as permanent as matter, energy, space, or anything else.

What is *not* permanent is something called time epoch, or a specific point in time. While time interval lasts indefinitely, time epochs, like yesterday afternoon or July 4, 1776, don't. Specific time coordinates are experienced only once in the present moment and then lost irretrievably to the past. We're constantly supplied with the new moments of which the present always consists, but these are savored only briefly and then lost as quickly as they're gained. It's this constant sense of loss of time epochs, coupled with our mortality, that lends time its unique attributes of urgency, vulnerability, and value.

It's often been speculated that the procession of time epochs might be reversible. If time is nothing more than that which is measured by change, the argument goes, then it's as subject to manipulation as are those changes. If change suddenly ceased, so would the flow of time. If the changes that define time could somehow be reversed wouldn't time flow backward, carrying us away from the future and toward the past? Common sense argues against such an outlandish proposal, but common sense also argued against the heliocentric universe of Copernicus.

The question merits serious consideration, for many changes by which we

■■■■■■■■■■■■■■■■■■■■■■■■■■■■■■■■■■■■■■

Precious Time

Because we value time so highly we're constantly devising gadgets, activities, and strategies to conserve, create, or stretch it. The most popular of these gimmicks is probably Daylight Savings Time.

Setting clocks ahead to gain daylight was first proposed—facetiously—by Benjamin Franklin in 1784. The idea wasn't considered seriously until the early 1900s, when it was seen as a way to reschedule activities so as to use less electrical power. It was initially put into practice during World War I, first by Germany, then by Great Britain and the United States.

Since then clocks have been set back and forth all over the globe in response to assorted protests and crises. They're usually reset by an hour but sometimes by two, usually on a national basis but sometimes on a state basis, sometimes on a year-round basis, and sometimes only in the summer.

Regardless of how it's employed, though, Daylight Savings Time doesn't really save any daylight. It merely enables us to better utilize the limited supply we have.

■■■■■■■■■■■■■■■■■■■■■■■■■■■■■■■■■■■■■■

mark time are, theoretically at least, quite reversible. There's no known reason, for example, why the universe couldn't just as easily contract as expand. The proponents of an oscillating universe, in fact, hold that just such a contraction will eventually take place. Similarly, the swirl of the Milky Way and the orbits of the planets about the sun and of satellites about their planets could all be run backward in perfect harmony with the laws of inertia and gravitation as we know them. The earth could just as well rotate from east to west and have western sunrises and eastern sunsets. Terrestrially speaking, Newton's three laws of motion are perfectly symmetrical with regard to time. Acceleration and deceleration are mirror images of one another and would operate without alteration were time to be reversed. The laws of electromagnetism and electromagnetic radiation are also, technically speaking, symmetrical with respect to time. None of these processes shows any clear preference for the direction in which they take place. All of this suggests that, while time runs in only one direction, that direction isn't necessarily inviolate.

But this fanciful speculation must stop here, for there are other laws that strictly prohibit the reversal of time. The most notable among these is the second law of thermodynamics. This law, you may recall, states that heat always flows from a warmer object to a cooler one, never the other way around. (See Unit II.15.) This seemingly innocent statement dictates the one-way flow of time and validates common sense.

Heat, remember, is ultimately the motion of molecules, and it's the collective enthusiasm with which molecules vibrate that determines their degree of heat, or temperature. As the degree of agitation increases so does temperature and vice versa. A hot spot is composed of molecules shaking relatively more excitedly than the molecules of a cold spot. As heat is transferred from the hot spot to the cold one according to the second law of thermodynamics, molecular motion is redistributed. Identifiable concentrations of lively and lazy molecules disappear. In their place arises a single, uniform pool of molecules all of which possess more or less average motion. A previous organization of molecules according to temperature has been lost. And once lost, this organization can never be spontaneously regained, for without some sort of external push heat never forsakes a warm spot to form cold and hot ones.

The second law of thermodynamics has implications that extend beyond the scope of heat and molecular motion. Any collection of sorted or systematized molecules, whether organized by temperature, size, density, mass, or type, will eventually revert into a disorganized, heterogeneous mixture if no effort from outside the collection is expanded to prevent them from doing so. This preference for uniformity to variety is called entropy. The second law of thermodynamics, broadly stated, says that the total amount of entropy, or disorder, in the universe is constantly increasing and that the amount of order is diminishing. Everything, starting on a molecular scale and proceeding on upward, moves toward randomness and away from structure, never the other way around.

The constant striving toward randomness has a few benefits. The atmosphere's oxygen, thanks to entropy, remains evenly distributed around the globe. No one with free access to the air will die of suffocation. And a spoonful of instant coffee, again thanks to entropy, dissolves equally throughout a cup of hot water. Each swallow tastes more or less like every other swallow, no matter how bad that might happen to be. The power of electricity is gained by harnessing the entropic urge of positive and negative charges as they seek to merge and cancel one another.

But the costs of entropy are enormous. Entropy dooms all efforts to organize to ultimate failure. Any process, in other words any change over a passage of time, eventually runs downhill, never the reverse. Entropy always increases, never decreases. And the universe grows only older, never younger, as time, like the events by which it's defined, runs only one way.

IV.3 | THE PAST AND THE FUTURE

One need not be a deep thinker or a trained scientist to realize that time always flows in one direction. There exists solid proof that even our primitive ancestors remembered a past and planned for a future. Cave drawings depicting heroic hunting scenes evidence the recall of time gone by and the wish to remember it. The formation of tools evidence the recognition of an impending time to come and the wish to prepare for it. It can be convincingly argued that a coherent sense of time, along with language and other manifestations of higher intelligence, has been in large part responsible for the survival of our otherwise poorly equipped species.

The earliest nomadic tribes probably had a very limited concept of the past and the future. Foresight undoubtedly extended little beyond the quantities of food or firewood that could be stockpiled, nor did it need to. The next kill was the most distant event that required planning. The past was limited to the span of memories directly experienced, with the oldest members of the clan having the greatest knowledge of bygone events. The concept of time in general was no larger than a human lifespan.

As hunting gave way to agriculture these thresholds of time's beginning and end expanded. The establishment of permanent residences and the refine-ment of language and writing allowed for the collection and retention of historical information stretching beyond memory. Oral history and legends were supplanted by written records, archives, and calendars. Collected data about eclipses; lunar, solar, and planetary motions; weather patterns; and the visitation of locusts and plagues were analyzed and drawn upon for guidance. The yearly cycle of planting and harvesting required long-range planning and a faith in distant time.

With the ownership of private property even the passing of generations took on added significance. Life was lived in the present, but that present was ever more firmly fixed within a broad context of time. Existence was more than a day-to-day affair; it was part of a long, unbroken span of human endeavor. If the here and now was to be maximized the there and then had to be taken into account.

The pushing back of the past and the extension of the future was restricted by certain practical matters, of course. Projections of the future then were subjected to all the uncertainties they are now. Beyond the approximate forecasts of seasonal weather changes visions of the future were an exercise of religion, mysticism, and superstition and in no way scientific. Documentation of the past was generally limited to the history of a given civilization. The Egyptians, the Babylonians, the Greeks, and the Romans each in turn assumed that the foundings of their own particular cultures marked the beginning of time and that anything predating them was inconsequential. Little serious thought was given to when the flow of time had its absolute beginning or end. And even if some dared consider such issues they could do no more than speculate, for the means with which to explore time were severely limited.

If the dimensions of time were at first approximately and informally restricted by a lack of tools, knowledge, and imagination, they were later more rigidly and formally restricted by the Church. Christian scholars, drawing upon geneological and other historical references in the Bible, carefully recon-structed a chronology of the entire world. The beginning of time, or creation, they concluded, occurred about 4000 B.C. The earth was then formed at the hand of God and later drastically altered during a great flood, which they dated at 2349 B.C. Since then, the Church maintained, little had changed, for God had created the world just as he had wanted it. It was destined to remain in its current condition until Judgment Day. This end of time, although not precisely fixed, was nonetheless expected within the not-so-distant future, possibly at any moment.

IV.4 THE INVESTIGATION OF TIME

It wasn't until the middle of the eighteenth century, some 400 years after the brilliance of the Renaissance first began to dispel the gloom of the Middle Ages, that the boundaries of time as set by the Church came under attack. The revolutionary who fired the first shot of this intellectual assault upon tradition was a French scientist, Georges-Louis Leclerc comte de Buffon (1707–1788). Like Copernicus and Galileo before him, Leclerc dared to challenge the authority of the Church and the weight of public opinion armed with only his own bold ideas and a bit of physical evidence.

Leclerc argued that the earth wasn't created out of nothingness at the hand of God in its current condition some 6,000 years ago, but that it was formed instead by physical forces and shaped to its present state by the passage of time. According to Leclerc, the earth's mass had originally been a part of the sun, had wrenched itself loose under the influence of centrifugal force, and had then been hurled into orbit where it cooled to a habitable temperature. To back up this line of reasoning he conducted a number of experiments in which he clocked the cooling times of heated spheres of various sizes, substances, and temperatures. Drawing upon the data thus assembled he estimated that the earth had been an independent body for something on the order of 75,000 years.

Due to a severe lack of theoretical and practical knowledge, Leclerc's calculations were laden with errors and produced a badly flawed solution. Nevertheless, he did multiply the accepted age of the planet by a factor of more than ten. Furthermore, he had breached a time barrier that had stood unquestioned, unassailed, and unaltered for hundreds of years. But perhaps even more significant was the fact that he had used scientific methods to explore the beginnings of time. It's Leclerc who gets credit for developing the powerful idea that present conditions, together with a knowledge of the workings of change, can be used to reconstruct the past.

The investigation of ancient history launched by Leclerc was taken up and intensified by a Scottish physician-turned-farmer-turned-geologist named James Hutton (1726–1797). By studying the geology of his native countryside Hutton formulated the doctrine known as uniformitarianism. In small words, this is the idea that changes happening in the present are examples of the same sorts of changes that have been proceeding at uniform rates for long spans of time. The salinization of the oceans, erosion, sedimentation, and the upheaval of geological plates taking place today, said Hutton, are the same things of which the physical past is composed. By measuring the rates of these changes and their total effect to date, the length of time that they've been under way can be determined. Hutton humbly refrained from making any guesses about the extent of the past or the future, preferring instead to say only that he could find "no vestige of a beginning, no prospect of an end."

The work of Hutton and a fellow Scotsman named Charles Lyell (1797–1875) established the investigation of time as a respectable scientific discipline. Uniformitarianism became a respected method of investigation, not only in geology, but in other areas as well. In biology it was uniformitarianism that helped Charles Darwin (1809–1882), a personal friend of Lyell, formulate his theory of evolution published as *On the Origin of Species by Means of Natural Selection* in 1859. Darwin, unlike Hutton, wasn't so timid about naming numbers. In order for his evolutionary process to unfold he required nothing less than hundreds of millions of years, an expanse of time hard to imagine and previously unheard of. With a single stroke Darwin not only erased the carefully guarded line separating man from the beasts, but also opened up a past of virtually unbounded dimensions.

Time explorers following Darwin found themselves in a predicament: the assumed scope of the past far outdistanced the tools and techniques available for its study. The sequence of fossils, from mollusks to amphibians to dinosaurs and on up through the mammals, could be logically arranged in relative order, but none could be absolutely dated in terms of years. Dinosaurs were older than mammals, but how much older no one could say. The geological formations predating the oldest fossils were even farther off limits. No perceptible contemporary change could be projected back over such enormous reaches of time with anything even resembling accuracy or reliability. It was like trying to trace out the orbit of Mars based on the sightings of a single night.

The problem seemed insurmountable until 1896, when Antoine-Henri Becquerel made his serendipitous discovery of radioactivity, the process whereby unstable atoms spontaneously degenerate. (See Unit I.26.) The radioactive decay of uranium-238 first noticed by Becquerel results in the transformation of uranium-238 into lead-206. Before the decay takes place there exists an atom of uranium-238. Afterward there exists an atom of lead-206. This transformation takes place in a very predictable manner and, more importantly, according to a very precise schedule. The decay of unstable isotopes is, in fact, a highly accurate clock ticking away inside certain types of matter.

Radioactive decay proceeds not in years or days, but in units of change called half-lives. Each unstable isotope has its own characteristic half-life by which it decays and marks off time. Half-lives range from the very short, like the .00016 seconds of polonium's least stable isotope, to the medium, like the 1,620 years of radium-226, to the very long, like the 4.5 billion years of

The Bible and Time

Although Christian scholars have often tried to use it to support their positions, the Bible is not a very good source of information when it comes to explaining the operation of the physical aspects of planet Earth. During the great heliocentric-geocentric controversy a few Bible passages were cited as proof of the geocentric cause. In the end, though, these passing references to the motion of the sun in the sky didn't have much effect on deciding whether or not the earth moved.

Biblical references to dates, while much more numerous, are equally vague, often confusing, and at times inconsistent. The inference of exact dates from Bible passages relies as much on guesswork and imagination as it does on hard facts. Certainly the most imaginative of all Christian chronologists was the Irish Archbishop James Ussher (1581–1656), who said the creation of the universe took place at precisely 9:30 A.M., October 23, 4004 B.C. One can only assume he meant Greenwich Mean Time.

Christ's date of birth has also been set with somewhat presumptuous precision. In fact, the Bible clearly states that Christ was born during the reign of King Herod the Great, which a variety of historical records fix as a period from 11 to 4 B.C. and not the year 1. (See Unit IV.9.) The date of December 25 is completely unsubstantiated. Indeed, certain evidence suggests that it may have its origins in ancient pagan celebrations of the winter solstice, with which it closely coincides.

Easter and the other religious holidays whose dates change from year to year are collectively called movable feasts. Their observation is tied to the moon, a custom that can be traced back to the Jewish lunar calendar. Easter is actually determined by the sun and the moon. According to an edict of the Council of Nicaea in A.D. 325, Easter is observed on the first Sunday after the first full moon following the vernal equinox.

—What are the earliest and latest possible calendar dates for Easter? Can we determine when Easter will fall in future years or must we wait for the moon?

—Is the historical accuracy of Christ's birth important? Why or why not?

—How do those who interpret the Bible literally explain the existence of fossils that scientists claim are millions of years old?

—Why do you think January 1 has been picked as the beginning of a new year?

uranium-238. During a single half-life of any radioactive isotope exactly one-half of its atoms degenerate into other substances. There's no way of knowing which atoms will decay and which won't or of knowing when any particular atom will self-destruct; nor is there any known explanation of why such a curious phenomenon takes place. The world's supply of radium-226 is halved every 1,620 years. Any isolated sample of radium-226 is similarly reduced according to the same schedule and replaced by other substances. Radioactive isotopes are rare for just this reason—they've long ago been reduced to extinction by the perpetual halving of their quantities. Uranium-238, with its extremely long half-life, is one of the few naturally occurring isotopes unstable enough to undergo decay yet stable enough to still be around in appreciable quantities.

When uranium-238 decays it leaves lead-206 in its place, atom for atom. Lead-206, conveniently, is not only stable but is produced only by the decay of uranium-238. Any encountered atom of lead-206 was, at one time, an atom of uranium-238. Any encountered atom of uranium-238 will be, at some time, an atom of lead-206. Lead-206 is never found in the absence of the uranium-238 from which it's being generated, nor is uranium-238 ever found in the absence of the lead-206 into which it's deteriorating.

The age of any uranium ore can be determined by measuring the ratio of its lead-206 to its uranium-238. A newly created rock would contain only uranium and no lead. An infinitely old rock would contain only lead and no uranium. Uranium ore samples mined from the earth's crust contain just about equal portions of uranium and lead. Approximately one-half of their original uranium content has decayed into lead. Just about one half-life of uranium-238, determined to be 4.5 billion years, has gone by since these rocks were new. Our planet, according to the radioactive dating of uranium-238 versus lead-206 as well as of several other parent-child isotope ratios, is about 4.5 billion years old.

IV.5 | THE BEGINNING OF TIME

If the world was really formed 4,500,000,000 years ago, then a number of other questions immediately come to mind. Questions like: Formed how? Formed out of what? As always seems to be the case, no matter what outrageous questions someone is willing to ask, someone else is willing to answer.

The first to push the origins of time back beyond the birth of our planet, of course, was Leclerc. By supposing that the earth had been flung off the sun like a clod of mud is flung off a spinning wheel, he supposed a universe that predated the globe. One of his countrymen took the idea a step further, to a universe that predated not only the earth, but the sun and the entire solar system.

In 1796 Pierre-Simon de Laplace (1749–1827) proposed what has come to be called the nebular hypothesis of the origin of the solar system. Because in his day the sun, every planet, and all of their satellites were thought to twirl in a similar direction, Laplace reasoned that the whole arrangement must have come from a common source and must have been formed during the same cosmic event. The source, according to the Frenchman, was a vast, diffuse cloud of interstellar dust and gas. The event was the contraction of this cloud under the influence of gravity. As the cloud closed in on itself, he theorized, it began to swirl like a whirlpool and then congeal into solid clumps. The outer, smaller clumps became the planets and moons. The central and largest clump became the sun.

This theory enjoyed a long popularity, surviving until about 1900, when certain mathematical models indicated that the distribution of angular momentum within the solar system was inconsistent with such a scenario. An American geologist, Thomas Chrowder Chamberlin (1843–1928), was ready with another theory to supplant the fallen one of Laplace. According to Chamberlin's planetesimal theory the sun came first. The rest of the solar system was spawned when another star sideswiped the sun, ripping away portions of it and flinging them into orbit. Like Leclerc's original speculations, this theory used the preexisting sun as the source of the planets, but ascribed their birth to an act of extraordinary cosmic violence rather than ordinary centrifugal force.

Again, statisticians eventually decided that such an interstellar fender bender was mathematically impossible. This allowed another astronomer, the German Carl Friedrich von Weizsacker (1912–), to revive the nebular hypothesis of Laplace, supported by more recent, more complete, and more favorable data. Since he did so in 1944 the nebular hypothesis, sometimes called the dust cloud hypothesis, has continued to be the most popular explanation of the birth of our planet, our moon, our sun, and the rest of the solar system.

If isotope ratios provide the answer to when the earth was formed, and the nebular hypothesis theory provides answers to out of what and how it was formed, then there obviously arises a need for similar answers about the cloud of dust that preceded it. This cloud, along with the rest of the matter in the universe, it's supposed, dates back to the absolute beginning of time. This inconceivably remote birthday hasn't been ignored, of course. It's been intensely investigated and duly approximated. The methods employed in this hunt for the first cause are all based on Hutton's doctrine of uniformitarianism.

The universe, according to all three of the most popular theories regarding its nature—the Big Bang theory, the oscillating theory, and the steady state theory—and fairly substantial physical evidence, is expanding. (See Unit III.17.) The expansion is attested to by the so-called red shift of light coming from the objects of distant space. The amount of red shift—the degree of reddening of a luminous body's light due to the Doppler effect—is directly related to the rate at which that body is receding from us. The greater the reddening the hastier the retreat. As discovered by an American astronomer,

■■■■■■■■■■■■■■■■■■■■■■■■■■■■■■■■■■

A Cosmic Year

Although the 17-billion-year age of the universe is impossible to imagine, the history of the cosmos can be illustrated by compressing it into a single year. As we look back at this cosmic year, then, it is now the stroke of midnight on New Year's Eve.

The universe was born with a Big Bang precisely one year ago. It wasn't until the end of April, though, that the Milky Way galaxy became a recognizable feature in this chaos of matter and energy. The solar system and planet Earth, meanwhile, didn't take form until the first half of September. It was already nearly Christmas—just a week ago—by the time dinosaurs evolved.

We humans didn't show up until early this morning, and have only been civilized for the last ten seconds or so. The Egyptians built the pyramids just eight seconds ago, the Greeks conquered Troy six seconds ago, the Romans ruled the Western world four seconds ago, and it has been just two seconds since feudalism prevailed in Europe. Each second equals a little over 500 years in this scheme, and a human life span is reduced to the blink of an eye.

■■■■■■■■■■■■■■■■■■■■■■■■■■■■■■■■■■

Edwin Powell Hubble (1889–1953), the rate of an object's retreat seems closely linked to its distance. The most distant objects are the ones moving away from us at the greatest velocities.

According to Hubble's law this is more than mere coincidence; the most remote stars are farthest from us precisely because they've been in a state of rapid withdrawal since the beginning of time. Other stars, traveling more slowly, have stayed closer. By plotting the course and speed of these diverging objects and then extrapolating the results backward into time via uniformitarianism a very interesting result is obtained: all courses converge, and do so more or less simultaneously. The convergence point is the site of the Big Bang, the cosmic egg from which everything hatched. The moment of convergence is the birth of the universe and the beginning of time. The date turns out to be a rather fantastic 17 billion years ago.

Other evidence seems to confirm this ultimate antiquity. The distribution of matter by element type within the observable universe closely matches that of a model based on a 17-billion-year-old cosmic egg and current ideas about the evolution of matter. The spatial distribution of galaxies also agrees with a 17-billion-year history. So does the distribution of galactic velocities. The distribution of stars according to age also fits into the same basic schedule. The amount of background thermal radiation in intergalactic space does, too. The universe, apparently, is 17 billion years old.

IV.6 THREE NATURAL CLOCKS

Compared to the exploration of time long since expired, the reckoning of current time would seen to be a relatively easy task. Such, unfortunately, is not the case. Time, even when it's immediately present, defies, or at least resists, any attempt to regulate, simplify, or organize it.

The problem lies in the fact that time operates according to its own unalterable and natural rhythms, and not according to any humanly devised or imposed standards. While volume, length, and mass can all be defined and arbitrarily chopped into pieces more or less as we wish in order to make them easier to manipulate and communicate, time can't. Time exists in certain preordained units and must be dealt with accordingly, in spite of all the complications, inconveniences, and inconsistencies that arise.

Time is meaningful to the extent that changes take place during its passage. It's through change that time is experienced and it's by regular, predictable change that time is measured. The most persistent, reliable, and notable changes available to measure time are the motions of the earth and the moon. Together with the sun as a reference point they operate to define the three basic, unalterable, and perpetual units of time upon which reality as we experience it operates: the day, the month, and the year.

The day, the alternation of dazzling sunlight and the dark of night, is the most obvious of these natural time units. Virtually every living creature on the face of the earth, from the humblest flora to the proudest fauna, recognizes and responds to this daily cycle. For timekeeping purposes the day is defined as the interval from one noon to the next at any given point on the globe. This period is properly called one solar day and consists of 24 hours, each of which is defined as being 1/24th of a solar day.

The month is the least apparent unit of natural time, especially in our modern age. Although spectacularly evident to any student of the skies and important to many ancient cultures, the lunar cycle has no significant impact on most earthly events and can and does go completely unnoticed by a large portion of the planet's human population. We rise every morning and retire every evening in conjunction with the sun, but very little, if any, of our activities are tied to the moon. Still, the natural month has played a major role in the attempts to keep track of time. A month, for timekeeping purposes, is defined as the interval from one full moon to the next. This period is properly called a synodic month and consists of 29 days, 12 hours, and 44 minutes, more or less.

Finally, there's the year. The passage of a year, though not acutely sensed as it happens, results in numerous cycles that rival the drama of many daily ones. Hibernations, procreations, migrations, and countless complete life cycles are attuned to the cadence of the seasons. Mankind, although apparently lacking a biological connection to the year, has come to appreciate its importance intellectually and has made large-scale adjustments to the yearly cycle of climatic change. A year, for timekeeping purposes, is defined as the interval from any given solstice or equinox to the succeeding solstice or equinox of the same type. This period is properly called a tropical year and consists of 365 days, 5 hours, and 48 minutes, more or less.

It's by the ticking of these three natural clocks—the solar day, the synodic month, and the tropical year—that time must be measured if it's going to be relevant to existence on planet Earth. At first glance there wouldn't seem to be much of a problem living according to this astronomically predetermined arrangement. Each basic unit is virtually constant, producing cycles of unfailing accuracy. Each is completely reliable and, according to Newton's three laws of motion, should remain so. Each is easy to observe and has been the object of careful scrutiny for thousands of years. Each is well understood and has been meticulously measured to the nth decimal place. Why, then, is the reckoning of time so cumbersome?

The answer lies within the very accuracy, dependability, and permanence of the basic time units nature has bestowed on us. Each is unalterable. Each demands strict allegiance. Yet each is totally independent of the other two and compatible with neither. The year can't be evenly divided into or conveniently expressed in terms of months or days. The month can't be evenly divided into or conveniently expressed in terms of days or years. And the day can't be used to define months or years. We've got three master clocks by which to gauge time, but no two of them agree. The logical reckoning of time, far from being straightforward, is a task fraught with profound and inherent difficulties.

IV.7 | THE LUNAR CALENDAR

The earliest, simplest human societies, of course, didn't really have all that much trouble keeping track of time, for they had very little to keep track of. Concerned neither with the distant future nor the remote past, they lived pretty much in the present moment. The day, in fact, fulfilled just about all their timekeeping needs. And the day, taken without any connection to either the month or the year, proved to be a fairly foolproof clock. The appearance of the sun in the morning clearly announced when it was time to get up, gather fuel, forage for edible treats, or mount a hunt. It's disappearance at night just as clearly signaled the time to seek shelter. The phases of the moon were interesting but had little practical importance. The passage of the seasons occurred according to a familiar sequence, but there was no recognition of the fact that this sequence followed a rigid schedule and no attempt was made to measure such a long span of time.

Needless to say, life eventually got more complicated. And as societies grew in size and complexity a more rigorous accounting of time became all but mandatory. It wasn't good enough just to know that spring followed winter anymore, it was necessary to know exactly when spring was due to arrive.

● ●

A Matter of Perspective

The lengths of the solar day, the synodic month, and the tropical year are all determined by astronomical phenomena as observed from earth. They're only relevant to us, and it's only by making ourselves a static reference point that they make any sense. No wonder Galileo had so much trouble convincing people that the earth moved.

There are other reference points, however, and other ways to measure these natural time units. Besides a solar day there is also something called a sidereal day. The word *sidereal* means "of the stars," and this day is accordingly measured from the vantage point of the so-called fixed stars rather than the moving earth. The tropical day is the time between successive noons at any point on the globe. A sidereal day, on the other hand, ignores the sun and measures a complete 360-degree spin of the globe in relationship to the stars. While the tropical day is defined as 24 hours, the sidereal day works out to something like 23 hours, 56 minutes, and four-and-a-fraction seconds.

There is also a sidereal month, measured in much the same way. While a synodic month states the time from one full moon to the next as seen from earth, a sidereal month ignores the earth and measures a full rotation of the moon on its axis relative to a reference point outside the solar system. For the curious, a sidereal month is 27 days, 7 hours, and 43 minutes, some 2 days, 5 hours shorter than its synodic counterpart.

Finally, there is a sidereal year as well. The tropical year measures the time between equinoxes, thereby relying on the tilt of the earth's axis as a reference point. The sidereal year, in contrast, measures the exact time of one orbit of the earth around the sun in relationship to the universe as a whole. These two years differ by about 20 minutes due to the fact that, like a spinning top, the earth wobbles a bit.

—The approximately four-minute difference between the solar and sidereal days, when multiplied by the number of days in a year, works out to 24 hours. Can you explain why?

—What so you suppose the cumulative difference between synodic and sidereal months works out to be over the course of a year?

—A consequence of the difference between the tropical and sidereal years can be noticed from earth. What is it?

—Do the sidereal time units have any importance whatsoever? To whom?

● ●

Besides the need to know when to plant cotton and rice there were yearly taxes to be collected, censuses to be taken, and elections to be held. The rise of organized religion also entailed the declaration and observance of special holy days that had to be kept track of. All in all, the unadorned day became entirely inadequate. Time had to be organized on a larger scale.

The first attempt we know of to come to grips with this large-scale organization of time happened in Mesopotamia some 5,000 years ago. There the Sumerians set up the first, understandably crude, calendar. The Sumerians, logically enough, turned from the day to the next-largest available natural time unit, the moon or month, in their struggle to put time on an orderly basis. The month, they decided, consisted of 30 days. After counting a couple hundred moons or so they also decided that 12 of them corresponded to one cycle of the seasons, or one year.

This was a nice concept, but not very workable. Because a synodic month doesn't exactly consist of 30 solar days, and because a tropical year doesn't exactly consist of either 12 synodic months or 360 solar days, the shortcomings of the Sumerian calendar soon became apparent. The months of the calendar quickly fell out of synchronization with the phases of the moon, and the year of the calendar eventually fell out of synchronization with the seasons of the year.

About a thousand years later, again in Mesopotamia, the Babylonians inherited this defective timekeeping device from the Sumerians. After convincing themselves that no amount of ritual, prayer, or sacrifice was going to alter the schedule of either the sun or the moon, they resigned themselves to altering the calendar. A true month, they realized, was actually less than 30 days long. Unfortunately, it was also more than 29 days long. To resolve the dilemma they used a string of months that alternated between 29 and 30 days. After some experience with this arrangement they found that it, too, failed to properly keep the calendar months and the natural moons in harmony. So, when things started to get a bit out of whack, they occasionally used back-to-back 30-day months to straighten them out.

The Babylonians, by alternating 29- and 30-day months and sticking in an extra 30-day month once in awhile, managed to keep the phases of the naturally driven moon and the months of the humanly contrived calendar running side by side. The synodic month could, with constant tinkering, be tamed. Their troubles weren't over, though. By committing themselves to 12 lunar months per year, the Babylonians also committed themselves to a year of 354 or sometimes 355 days. Being good astronomers, they recognized that their lunar-reckoned year wasn't the same as the year according to the seasons, the sun, and the stars. Ever flexible, they solved this problem by inserting an extra, unlucky, thirteenth month into some years.

The Babylonian calendar required constant attention and tuning. The intercalation of extra days into some months and extra months into some years kept it aligned with the moon and the sun, respectively. But this same intercalation with which they purchased accuracy cost them dearly in terms of

■■■■■■■■■■■■■■■■■■■■■■■■■■■■■■■■■■■

The Metonic Cycle

The discovery of a pattern of intercalation in the lunar calendar is credited to a fifth-century B.C. Athenian astronomer named Meton. The Metonic cycle, like all calendar schemes devised before or since, is based on sheer coincidence rather than any underlying natural principle, in this case the fact that 19 tropical years contain almost exactly 235 synodic months.

There is, however, a two-hour disparity between the two cycles, so Meton's calendar errs by some 6 minutes a year. This is better than the Julian calendar, used for a millennium and a half, which is accurate to only about 12 minutes a year. The best thing we've been able to come up with, our current Gregorian calendar, still wanders by some 26 seconds per year. (See Unit IV.10.)

The biggest problem with the Metonic cycle isn't its accuracy, but its complexity. While the Gregorian calendar requires only the intercalation of a single extra day every 4 years, Meton's requires that numerous months contain extra days and that 7 out of every 19 years contain a whole extra month.

■■■■■■■■■■■■■■■■■■■■■■■■■■■■■■■■■■■

reliability. Because extra days and months were added purely on an as-needed basis and not according to any regular schedule, the calendar, precise in the moment, couldn't be confidently projected into the future. No one knew for sure which months would have 30 days and which would have 29, nor which years would have 12 months and which 13, until the astronomical events to which they were tied unfolded.

By the time the Babylonian culture was eclipsed by that of the Greeks, a long and detailed written record of the lunar calendar's performance had been accumulated. This history, like anything else that fell into Greek hands, was thoroughly analyzed. The result was the discovery of a pattern of intercalation that repeated itself every 19 years. By projecting this pattern into the future, extra months and days could be scheduled at prearranged intervals rather than wedged in every once in awhile. This development resulted in a lunar calendar that showed considerable promise of providing not only current accuracy, but long-term reliability as well.

IV.8 | THE JULIAN CALENDAR

When the Romans assumed control of Western civilization they rather casually ignored this rather sophisticated version of the lunar calendar in favor of one they had already developed on their own. Theirs, too, was a lunar device whose basic unit was a moon, or month. Each Roman month was elaborately organized by calends, ides, and nones. The calends marked the start of each month according to the new moon. The ides denoted the midpoint of each month according to the full moon, and midway between the calends and ides were nones.

But, while lavishing attention on detail, the Romans all but ignored the big picture. What they ended up with was an abbreviated 10-month, 304-day cycle that barely qualified to even be called a year. The month of October, for example, originally scheduled for fall, was in summer the next year and in spring a couple of years later. To remedy the situation two additional months were hastily inserted and the year extended to 355 days. The repair job was rather shoddily done, however. One of their months ended up with 28 days, seven of them with 29 days, and the remaining four with 31 days. What's more, the two new months were jammed in not at the end of the year, but right in the middle. What were formerly months number seven, eight, nine, and ten, and accordingly named September, October, November, and December, retained their old, now-nonsensical names when they assumed their new positions.

Despite this major overhaul, the Roman calendar still had all the problems to be expected of a 355-day year. To keep it more or less in step with the seasons it became necessary to intercalate months of 22 or 23 days every other year. But this proved to be virtually impossible, for the management of the calendar, rather than being entrusted to priests or astronomers, was instead placed under the jurisdiction of the politicians. They had every right, which they exercised liberally, to meddle with it, thereby extending their own terms of office, curtailing those of their adversaries, accelerating the collection of taxes, or otherwise manipulating schedules and events.

This chaos of chronology continued until the reign of Julius Caesar (100–44 B.C.). In what we reckon now as 47 B.C. Caesar made an eventful trip to Egypt. There he met not only Cleopatra, but Sosigenes (1st century B.C.), a Greek-Egyptian astronomer. Sosigenes related to Caesar how the Egyptians had pretty much abandoned the lunar calendar in favor of a solar one and suggested that the Romans do the same.

The Egyptians, just like the Sumerians, the Babylonians, the Greeks, and the Romans, had at first succumbed to the lure of the moon when they began to track time beyond the span of several days' worth. But the moon, no matter how closely calibrated or how carefully charted, consistently failed to predict the annual flooding of the Nile. This, their most important annual event, would occur one year during a full moon, the next during new moon or a first quarter. There was no discernible pattern. But, they eventually noticed, the first appearance above the horizon of Sirius, the Dog Star, occurred just before the flooding every year. And so they started to count the days between Sirius's arrival, thereby using a stellar rather than a lunar calendar.

Eventually they came to associate the movement of Sirius with that of another more obvious and easier-to-track star—the sun. They were then ready to develop the first truly solar calendar. It consisted of 365 days every year and totally ignored the start, end, and number of lunar cycles that it happened to include. Calendar units called months were retained for certain religious purposes and out of habit, but were reduced to mere arbitrary divisions of the year and in no way whatsoever connected to the comings and goings of the moon.

Caesar returned to Rome not only smitten by Cleopatra, but inspired by Sosigenes. Following the advice of the wise astronomer, he put Rome on a solar calendar, where Western civilization has remained ever since. In order to rectify past mistakes he extended the first solar year, what we now call 46 B.C. and what was then called the year of confusion, from its normal 355 days to 445 days by the insertion of 23 extra days at the end of February and 67 more between November and December. Succeeding years had 365 days each and were organized into 12 unchanging, nonlunar months. Futhermore, enacting a strategy that the Egyptians had considered but never used, Caesar decreed that every fourth year would be a leap year and would consist of 366 rather than 365 days.

This so-called Julian calendar closely resembled the one used today throughout the world. The sun, not the moon, dictated the passage of large-scale time and the length of the year. Three 365-day years, together with a 366-day leap year every fourth year, resulted in an average year of 365.25 days, a close approximation of a true tropical year.

IV.9 | THE NUMBERING OF YEARS

The Julian calendar endured the passage of time much better than did the empire that popularized it. For centuries following the fall of Rome the year remained at 365.25 days and the calendar marched along in step with the seasons. The 12 months, divorced from the moon in all but name and origin, remained around 30 days each and completed their designated cycle once a year. Farmers, politicians, astronomers, merchants, and scholars were all fairly content with the arrangement.

Only the Church had complaints. While it recognized the value of Caesar's device as a timekeeper and a coordinator of worldly activities, it was disappointed with the calendar's neglect of spiritual matters. The Julian calendar, with its pre-Christian origins, didn't pay due respect to religion. The power vacuum of the Middle Ages provided the Vatican with plenty of opportunity to remedy this deficiency.

The first adjustment was the adoption of the week to replace the calends, the ides, and the nones, whose lunar ties were viewed as being a bit pagan. The week, or something resembling it, had historically existed in nearly every organized society, although informally. A cycle of anywhere from four to ten days proved to be convenient for setting market days, days of rest, and days of worship. The Church, following the passage in Genesis that says that God rested on the seventh day of the creation, decided on a seven-day week. Now, besides the mathematically irreconcilable cycles of solar days, lunar months, and tropical years there was also the Christian week.

The Church also took issue with the way the years were numbered. After the grouping of days into months and months into years, the stringing together of a number of years had been the next logical step. This large-scale organization of time required that each individual year in a series be positively identified. When the Sumerians set up the first calendar they did this very informally. Special years were remembered by the major events that took place within them—an emperor's death, a military triumph, or a drought. Other, less notable years, were identified by their relationship to these landmarks.

The Greeks formalized this technique, naming each year after the governor then in office and numbering each year within his term. This resulted in a logical and clear identity for any given year. There were a few problems, though. To correctly sequence the years one had to know all the governors. And, seeing as how each state had a different governor whose terms didn't always coincide, each year had a different name and number depending upon who was doing the labeling.

It was the Romans who put the numbering of years on a truly solid basis. They used the supposed date of the founding of Rome, what we would call 753 B.C., as the first year, or 1 A.U.C., for *ab urbe condita,* "from the founding of the city." All succeeding years were numbered consecutively from that point onward. Julius Caesar initiated his calendar reform in the year 709 A.U.C., 709 years after the founding of Rome. This numbering system was pretty much foolproof and remained in use into the Middle Ages along with the Julian calendar, with which it was closely associated. No one really cared when Rome was founded anymore, but it served as well as any other date, had the weight of custom to support it, and provided some much-needed standardization.

The Church, however, had other ideas. In about the year 1280 A.U.C. (A.D. 526) the Abbot of Rome, Dionysius Exiguus (c.500–c.560), proclaimed that the birth of Christ, not of Rome, should be the critical moment to which the clock of history should be set. According to this monk's calculations the hallowed event took place on December 25, 753 A.U.C. Using this as his starting point he proclaimed that the following year, what had up until then been 754 A.U.C., would henceforth become A.D. 1, for *anno Domini,* "in the year of the Lord." All subsequent years were to be renumbered to conform with the new starting point. The years preceding Christ's birth, said the Abbot, were to be counted backward from that point and labelled B.C., for "before Christ."

The fancy B.C./A.D. designations, along with all the tedious conversions they entailed, didn't catch on immediately but have, over time, become as close to a world standard as anything gets. Even non-Christian countries reluctantly use it for international purposes. The Jews, for instance, insist that the world was created in 3761 B.C. and have their own calendar, much like the lunar one

■■■■■■■■■■■■■■■■■■■■■■■■■■■■■■■■■■■■■

Chronological Quirks

Although our current method of chronology is more than adequate in most regards, it's not without its peculiarities. Most of these result from the fact that the scheme consists of two separate numerical sequences, one of which runs backward.

This numerical redundancy is most troublesome for the years nearest zero, which could easily be either B.C. or A.D. and so need to be carefully labeled. There is, by the way, no year zero. This needs to be taken into account when figuring time spans that include the transition from B.C. to A.D. The period 10 B.C.–A.D. 10, for example, consists of 19 years, not 20.

Because the first year is number 1, the first century consists of the years 1–100, the second century the years 101–200, the third century the years 201–300, etc. This, by extension, is what makes the 1800s the nineteenth century, the 1900s the twentieth, etc. Notice that all centuries begin with a year ending in the digit 1. Technically, the twenty-first century (or the third millennium) will start on January 1, 2001, not January 1, 2000.

■■■■■■■■■■■■■■■■■■■■■■■■■■■■■■■■■■■■■

used by the Greeks, which is dated from that time. For dealing with the rest of the world, though, they follow the Abbot with one qualification: instead of A.D. they use C.E. for "of the common era," and instead of B.C. they use B.C.E. for "before the common era."

Likewise, the Muslims have two calendars, the familiar Christian one for external use and one of their own design for internal use. The Islamic calendar uses Muhammad's flight from Mecca, in what is generally denoted as A.D. 622, as its first year. Even more interestingly, this calendar is a purely lunar one, much like that used by the Babylonians. Consisting of an unchanging 354 days, broken down into 12 months of alternating 29- and 30-day durations, it has no connection to the sun or the seasons. Used without any intercalation whatsoever, the lunar calendar of the Muslims completes almost 33 yearly cycles every 32 tropical years.

IV.10 | THE GREGORIAN CALENDAR

The Julian calendar, embellished with the seven-day week and reset to the birth of Christ, continued to be widely used without any further modifications up until the sixteenth century. By the early 1500s, however, a built-in flaw in its design began to take on significant proportions.

The pattern of three 365-day years followed by one 366-day year produced an unerring progression of calendar years with an average length of 365.25 days, or 365 days, six hours. Meanwhile, just as unerringly, the tropical years were ticking away at a pace of 365 days, five hours, 48 and a fraction minutes. Each year the calendar took some 12 minutes longer to get from one January 1 to the next than the earth took to get from one vernal equinox to the next. The 12-minute discrepancy, negligible for any single year or even any hundred years perhaps, mounted up over the centuries. The calendar set into motion by Julius Caesar in 46 B.C. was, after a run of 1,500 years, about 10 full days out of kilter with the sun, the solstices, and the seasons.

While this didn't cause any widespread panic among the general populace, it was an issue that caused some concern in learned circles. It was left up to the Church, the nearest thing there was to an international regulatory agency in those days, to do something about it. The Church's role, it should be pointed out, wasn't totally motivated by civic duty. If things were allowed to run their course, Christmas, dictated by the calendar, and Easter, dictated by the spring equinox, were certain to eventually be celebrated on the same weekend.

The first attempt to straighten out this sticky mess was undertaken by Pope Leo X in 1514. His efforts ended up being totally fruitless, and are of no special interest except for the fact that one of the astronomers consulted on this occasion was Nicholas Copernicus. It was upon considering the issue of calendar reform that Copernicus realized that the whole heavenly apparatus upon which it was based needed to be better understood. Copernicus rejected the Pope's request for assistance in rearranging the calendar, focused his energies elsewhere, and ended up rearranging the universe instead. (See Unit III.4.)

A half-century later Pope Gregory XIII (1502–1585) succeeded in instituting the much-needed corrections. A team of experts, headed by the German mathematician Christoph Clavius (1537–1612), worked on the technical aspects of the solution for almost ten years. Church authorities wrangled over the practical aspects of implementing them for another six. The new Gregorian calendar received final approval in 1582 and was put into operation the next year. This entailed one current adjustment to realign the date with the seasons and a second adjustment to prevent, or postpone really, the need for future adjustments.

To compensate for the 10-day drift of the Julian calendar over the previous 1,500 years it was necessary to chop 10 calendar days from the year 1583. The days eliminated were October 5 through October 14. In 1583, another, more modern year of confusion, October 4 was immediately followed by October 15. Then, to better synchronize the calendar and tropical years, it was decided that all future centenary years not evenly divisible by 400 would no longer be leap years as scheduled by the Julian calendar. The year 1600, divisible by 400, was thus a leap year just as it would have been without the reform. The year 2000 will be, too. The years 1700, 1800, and 1900, however, weren't leap years, nor will be the year 2100. This subtle change is all the

■■■■■■■■■■■■■■■■■■■■■■■■■■■■■■■■■■■■

Two Calendars

The replacement of the traditional Julian calendar with its more accurate Gregorian counterpart, initiated in 1583, took over 300 years to complete. The last country to make the switch was Russia, and it wasn't until 1918, after the Bolshevik Revolution finally dislodged the grip of the Eastern Orthodox Church, that it did so. By that time the gap between the calendars had grown to 13 days. Rather ironically, with the new dating system in place the anniversary of that October revolution is now observed on November 7.

England and its colonies, including what is now the United States, made the conversion in 1752. An additional source of confusion up until that event was the fact that prior to 1752 England had used March 25 rather than January 1 as New Year's Day. English dates in January, February, and most of March thus differed from their Gregorian counterparts not only by 11 days, but by a year as well. George Washington's birthday, for example, could be correctly cited as either February 22, 1732 N.S., or February 11, 1731 O.S.

■■■■■■■■■■■■■■■■■■■■■■■■■■■■■■■■■■■■

difference there is between the Julian and Gregorian calendars. It seems slight, but while the Julian calendar slips a day out of whack with the sun every 130 years or so, the Gregorian calendar is accurate to within one day every 3,323 years, and compensations for even this slight error have already been arranged.

The proclamation of Pope Gregory XIII was obediently accepted by the Catholic countries. They all set their calendars ahead 10 days as instructed. In the Protestant countries, however, the newfangled calendar was received with a bit less enthusiasm. Still basking in the glow of the recently successful Reformation, they weren't about to cave in to any Papal edict even if it did make sense. They stubbornly clung to the Julian calendar in its traditional defective condition. There were, then, starting in the month of October 1583, two widely accepted calendars marching side by side but out of step by 10 days. What was October 15, 1583, in Spain was October 5, 1583, in Holland. To make the chaos somewhat manageable the dates according to the Gregorian calendar were designated as N.S., for new style, while those according to the Julian calendar were designated O.S., for old style.

This absurd state of affairs continued for some time, worsening in the process. After the year 1700, when the Julians observed their regular leap year but the Gregorians didn't, the gap between them grew to 11 days. But eventually, recognizing their stubborn folly and swallowing their pride, the non-Catholic countries, one by one, came to accept the Gregorian calendar and make the necessary adjustments. With the design and universal acceptance of the Gregorian calendar the keeping of long-term, sometimes called big, time has at last been reduced to a straightforward, reliable, standardized, and, for all practical purposes, exact formality.

IV.11 | TIMEKEEPING

The year stands alone as the natural unit with which we organize, measure, and compile expanses of time beyond our immediate grasp. The day, on the other hand, is the natural unit we use to organize, describe, and sense the present moment as well as the immediate past and future. The day, like the year, is a basic reference point from which other units of time are derived. Just as the year multiplied becomes a decade and divided becomes a month, so the day multiplied becomes a week and divided becomes an hour. It's this process of multiplication and division, as well as the recording and cataloging of the resultant units, with which timekeepers have historically busied themselves.

So-called small time has been the subject of these efforts ever since human society developed to the point where various activities had to be coordinated within the span of a single day. Again it was the Sumerians who were first. Few specifics are known about how they dealt with small time, but the Sumerians, evidently to parallel the division of the year into 12 lunar months, divided the day into 12 smaller segments in order to make the identification of time within any single day more exact.

The Egyptians, following the Sumerian lead, probably used the same rationale sometime later when it was their turn to deal with time on a current basis. Actually, it wasn't the day, but the night, which they first broke down into 12 pieces. Each division was marked by the appearance of a certain star or group of stars above the horizon and observed by the performance of religious ceremonies.

Eventually, being symmetrically minded, the Egyptians decided that the day as well as the night should be split into 12 fractions. Each full day, from sunrise to sunrise, therefore contained 24 separate and distinct pieces—12 during the dark hours and 12 during the light ones. But because there was only a single star, the sun, visible during the daylight hours the procedure for measuring these divisions was necessarily quite a bit different. Instead of noting the appearances of multiple stars it was the movement of the solitary sun by which the day was marked off.

To accomplish the division the Egyptians used a shadow clock, which with later refinements became the sundial. As the sun rose in the east, arced across the southern sky, then set in the west the shadows it cast pointed first west, then north, then east. The shadow from a central post or gnomon sweeping across the face of the sundial graphically displayed the progress of the sun and the aging of the day. The Egyptians, then, can be credited with the creation of the 24-hour day and with the invention of the first device with which to gauge it.

The Egyptian arrangement had a few drawbacks, of course. For one thing, it relied upon an unobstructed visibility of the heavenly bodies. This made telling time an exclusively outdoor activity and placed it at the mercy of the weather. For another thing, it used sunrise and sunset as its set points and these varied throughout the course of a year. As the year progressed toward the summer solstice, the daylight portion of the day and each of its 12 fractions expanded while the night and its fractions shrank. Once past the solstice the reverse occurred. Perhaps it was because of these limitations that timekeeping in Egypt was never used for anything much beyond religious purposes.

Nevertheless, the ideas of breaking time down into units smaller than a day and using a clock to do so were destined to have lasting and powerful effects on the course of human development. Once embedded in the common consciousness, small time began to take on a significance of its own apart from the celestial events by which it was defined. Clocks became more important, stars less so. Time assumed a functional rather than a ceremonial significance. This domestication of small time required that it be taken out of the hands of a chosen few holy men and put it into the average home. It required the development of uniformly sized hours and affordable clocks by which to enumerate them.

The Greeks attacked this assignment with typical imagination and zeal, inventing the candle clock, the sand clock, and the water clock. But even though the Greeks effectively brought small time indoors and put it at the disposal of the average citizen, they left a few problems unsolved. The candle, keeping time as it burned down from one designated point to the next, eventually went out. The sand clock, still sometimes encountered as an egg timer, eventually ran out of sand. The water clock, or clepsydra, eventually ran out of water or clogged up. All required frequent maintenance. There was also the problem of quality control. No two candles, no two hourglasses, and no two clepsydras ever agreed with anything more than a general approximation. Each worked independently and kept its own individual time. And none ran with anything approaching fidelity to the sun or stars, the celestial master clocks.

The twin problems of reliability and accuracy continued to vex clock-makers for centuries and to some extent still do. The sundial, the hourglass, the candle, and the clepsydra were all tinkered with, refined, and elaborated upon to the limits of available technology, workmanship, and materials, but the effectiveness of each remained strictly contained by the weaknesses inherent in its design. What was needed was a fresh approach.

This was provided sometime early in the fourteenth century with the appearance of the mechanical clock. While its forerunners each worked by subdividing a long, slow motion—the movement of the sun, the combustion of candle wax, the flow of sand or water—the mechanical clock worked by multiplying a series of short, rapid motions. It was this switch from division to multiplication, from continuous to periodic motion, that really signified the beginning of the modern era of timekeeping.

IV.12 | SUPREME ACCURACY

The first mechanical clocks, whose inventors and builders remain mostly anonymous, were elaborate contraptions of gears, wheels, pulleys, and levers. All had two things in common. To begin with, all were powered by the force of gravity. The propulsion of gravity was captured by the use of falling weights. These weights were attached to long cords or chains. The cords or chains were, in turn, wrapped around drums or barrels. As the weights fell under the pull of gravity the cords unwound, thereby turning the drums and setting the entire clockwork into motion.

Secondly, all depended upon a device called an escapement to transform the otherwise rapid free-fall of the weights into a series of short, slow descents. The escapement, still used, is like a delicately balanced teeter-totter, dipping from side to side. As one side dips it locks the works of the clock by engaging a toothed wheel, thereby stopping momentarily the fall of the weights. The pressure of the suspended weight then kicks the escapement the other way, releasing the clockwork in the process. The falling weight then kicks the escapement back again into its locked position. This regular engaging and disengaging of its apparatus allows the clock to advance only one tooth's worth at a time, and it's the flip-flopping of the escapement that produces the classic tick-tock sound of mechanical clocks.

The construction of mechanical clocks rapidly became something of a mania. All across Europe neighboring towns competed with one another to see which could build the most elaborate, most ornate clock tower. An integral feature of all these clocks was a system of audio or visual effects to announce the arrival of each new hour. Massive gongs, melodic carillons, parading apostles, signs of the zodiac, flying angels, revolving stages, and cannon fire were all used. Derived from either the German word *glocke* or the French word *cloche,* both of which mean "bell," the word *clock* evidently dates from this period. The pocketwatch, too, became popular around this time when someone discovered that the escapement could be powered not only by a falling weight but by a coiled spring.

Despite the fascination with clocks and watches, though, timekeeping itself remained more of an art than a science. Each timepiece was handcrafted and had its own idiosyncrasies due to imperfections of design or workmanship. All were highly susceptible to variations due to changing weather conditions or years of use and wear. More importantly, the escapement itself had no natural time sense, but merely reacted to the forces applied to it. Accuracy could be sought only via trial and error, constant adjustments, or dumb luck. A clock that neither gained nor lost more than 15 minutes a day was considered exceptional. Standards for watches were even lower.

This technological impasse was shattered in 1582 while a 20-year-old medical student sat in the cathedral in Pisa, Italy, distractedly watching an overhead lamp sway to and fro on the end of the chain from which it was suspended. The student, named Galileo, using his pulse to confirm his observations, noted that each swing of the lamp took the same amount of time, even as the size of the swings shrank or grew. Subsequent experiments confirmed that any given pendulum maintains a constant periodic motion regardless of the size of the arc of its swing or the weight of its bob. Like the earth's orbit and rotation, and unlike the escapement, the pendulum is a natural timekeeper with a built-in sense of rhythm.

A new generation of clockmakers quickly exploited this breakthrough to produce clocks of unprecedented accuracy. The most notable among them was Christian Huygens (1629–1695), a Dutch astronomer. He succeeded in attaching a pendulum to an escapement and thereby produced an instrument that kept time to within a few minutes a week. The escapement, married to the pendulum, was no longer a victim of circumstances, but a true dictator of time. Huygens also developed the balance spring, a delicate band of steel that did for watches what the pendulum did for clocks. Together the pendulum and the balance spring ushered in a whole new era in the art of timekeeping.

The improvement in accuracy resulted not only in the more precise determination of the hour, but also in the need for even smaller units of time. So the hour was divided into minutes and minutes in turn divided into seconds. Eventually, of course, like the day and the hour, even minutes and seconds became too crude to measure small time satisfactorily. Science and industry especially needed to shave time into even smaller and more accurate slivers. This they have managed to do by improving upon the principle of the pendulum—by finding ever more rapid natural rhythms.

Thanks to their remarkable success in this endeavor, time, despite being no better understood than it was a thousand years ago, is now the most easily measured physical commodity known. There now exist clocks that can calibrate the time it takes a ray of light to travel from one side of an atom to the other. There now exist clocks that keep time to within a few seconds every billion years. There now exist, in short, clocks capable of supreme accuracy—clocks capable of doing just about anything a clock can be asked to do.

IV.13 | IRRECONCILABLE STANDARDS

As the measurement of small time, very small time, and very, very small time became ever more exact, its basic reference point remained unchanged. All units of small time, from the hour on down to the thinnest sliver of a second, were defined in terms of the solar day. An hour was defined as 1/24th of a solar day; a minute as 1/60th of an hour, or 1/1,440th of a solar day; a second as 1/60th of a minute, or 1/86,400th of a solar day; and so on down the line.

For many years there were no problems with this system, which used multiples of very small periodic motions like tuning fork vibrations to measure time and divisions of a long periodic motion—the solar day—to define it.

Faster and Faster

All our efforts to measure time are based on the existence of natural periodic motions. Greater accuracy can only be obtained with a faster standard of measurement. The day was the obvious starting point. It wasn't really improved upon until Galileo discovered the much more rapid regularity of the pendulum. A pendulum can be made to swing faster by shortening it, but this strategy can be effectively pursued only up to a point. The next breakthrough in timekeeping required a new and faster natural rhythm.

Such a rhythm was found in the tuning fork, a two-pronged instrument that vibrates when struck to produce a musical tone of remarkable clarity and unwavering pitch. The pitch of a tuning fork is determined by the frequency at which it vibrates, and this frequency, like the rotation of the planet or the swing of a pendulum, has a built-in sense of time. But, while the planet spins once a day and a pendulum swings maybe once a second or so, a tuning fork vibrates hundreds of times a second.

The frequency of a tuning fork has its limits as well, of course, so subsequent advances in chronology had to wait until the discovery of even faster natural rhythms. A quartz crystal subjected to a tiny dose of electric current was found to oscillate regularly at a frequency of around 100,000 times a second. Then it was discovered that a molecule of ammonia, stimulated by certain electromagnetic radiation, vibrates regularly and at a rate of millions of times per second. A cesium atom, similarly energized, ups the pace into the billions. A hydrogen atom vibrates faster yet. These so-called atomic clocks measure time in units of millionths of seconds, called microseconds; thousandths of microseconds, called nanoseconds; and thousandths of nanoseconds, called picoseconds.

—What additional demands were made upon timekeepers with the development of coast-to-coast radio and television broadcasting?

—What are the advantages and disadvantages of an electric clock? What is the source of its regularity?

—How many kinds of clocks and watches do you see during the course of a day? Can you figure out what natural rhythms they rely on to keep time?

—Why do you suppose hours and minutes are broken down into 60 smaller units rather than 100, which seems more logical?

Eventually, though, things came to a head. The second as measured by tuning fork vibrations, it seemed, didn't always agree with the second that was a certain fraction of a solar day.

The reason for the occasional incongruity of these two different seconds, based on separate standards, was a simple one: while all vibrations of a tuning fork or all oscillations of a quartz crystal are equal and unchanging, all solar days aren't. The earth, it turns out, rotates at a slightly variable rate as it travels at slightly different speeds at slightly different distances from the sun along its slightly elliptical orbit. The solar day, along with all the parts into which it can be divided, contracts and expands a bit during the course of a year.

This fact, once discovered, was intolerable to the keepers and users of small time. There was simply no way a constantly changing value could be used as a standard. So the official second was redefined—not as 1/86,400th of just any solar day, but as 1/86,400th of an average solar day. The average solar day, in turn, was defined as a tropical year divided by the number of solar days it contained. This new definition was based on the assumption that while the length of a day may ebb and flow a bit during the course of a year, these fluctuations would all eventually cancel out.

The new definition of the second, as well the assumptions upon which it was based, held up for only a century or so before more advanced technology, more complete data, and a more thorough understanding of natural time cycles rendered it obsolete. What signaled its doom was the discovery that the solar day didn't only shrink and expand during the course of a particular year, but that the length of an average solar day was steadily lengthening as the years mounted. The explanation for this growth of the average solar day is the only one possible: the earth's rotation on its axis is slowing. This slowing is apparently due to tidal friction, the constant sloshing of sea water against the underwater portions of the earth's crust.

Because of this long-term slowdown the definition of the second based on an average solar day is untenable, for the average solar day of any year is slightly longer than that of the preceding year. To once again put small time on a solid foundation it was necessary, in 1956, to redefine the second as 1/31,556,925.9747th of the tropical year that began at noon on January 1, 1900. It has since, mostly for the sake of convenience, been restated as 9,192,631,770 cycles of the radiation given off by a cesium atom under specified conditions.

Despite, or maybe because of, all this hairsplitting, there exist two quite distinct yet equally valid standards of small time: the cesium atom second and the solar day. The cesium atom second is used by physicists and technicians. The solar day is used by navigators, astronomers, and calendar makers. The two are fundamentally irreconcilable. Neither can be properly expressed, measured, or defined in terms of the other, for their relationship constantly changes as the earth slows down.

The situation bears a close resemblance to big time's quandary. Here as well there are two equally valid standards: the solar day and the tropical year. Each of these also defies definition in terms of the other, for they operate independently of one another according to irreconcilable schedules. There's simply no way to synchronize the solar day with the tropical year for the keeping of big time, nor is there any way to synchronize it with the cesium atom second for measuring small time. In each case the separate standards must be allowed to go their own ways and then be periodically fine-tuned to one another by the ancient process of intercalation first used by the Babylonians.

Where big time is concerned the intercalation takes the form of the insertion of an extra day into leap years. Where small time is concerned the intercalation takes the form of the insertion of an extra second into "leap days." These seconds are customarily sneaked in at midnight on December 31. Under the direction of the International Earth Rotation Service, certain years, on an as-needed basis, contain a one-second void between 11:59:60 P.M. on New Year's Eve and 12:00:00 A.M. on New Year's Day, during which the official atomic clocks pause to let a slowing earth get back into step.

IV.14 | STANDARD TIME

Through constant vigilance and intermittent adjustment we manage to keep the cesium atom second, the solar day, and the tropical year marching in what is, for all practical purposes, perfect step. Small time, big time, and time interval in general is, as far as we can see, about as well figured out and regimented as it's ever going to be. The question How long?, whether it be in terms of nanoseconds or centuries, can be answered decisively and authoritatively.

The question When? can also be satisfactorily answered. But determining a point in time, or time epoch, is an altogether different assignment with altogether different problems. These problems all stem one way or another from the fact that time, even when measured with sophisticated chronometers and calibrated to a hundred decimal places, is, was, and always will be a totally unbroken and unbounded commodity.

Despite our efforts to chop it up into convenient pieces, time itself recognizes no divisions. There's no natural demarcation of any kind that signals the beginning or end of any second, day, or year. And, despite our efforts to affix it to the founding of Rome or the birth of Christ, time contains no natural reference points. It may have had a beginning and may also have an end, but in the meantime it simply surrounds us with a perpetual and pervasive now.

In truth, there's really no way to determine our exact location on this uncharted, uninterrupted, and shoreless sea of time. The best we can do is fix our position in relationship to something familiar. This familiar something, once again, is the sun. Just as units of time duration, like the day, depend upon our cumulative motion relative to this benevolent blast furnace, so points in time, like high noon, depend upon our current position relative to it. The time

■■■■■■■■■■■■■■■■■■■■■■■■■■■■■■■■■■■■

The 24-Hour Clock

The use of two sequences of hours every day, each identically numbered from 1 through 12, is inherently confusing. Indeed, noon and midnight are hard to keep straight even when properly identified by the A.M./P.M. designation this system requires. Because the day has 24 hours it would only seem logical that the clock should, too.

Some clocks do. In order to avoid ambiguity, much of Europe, the United States armed forces, and numerous communication and transportation agencies around the world use a clock that completes a single 24-hour cycle each day.

The morning hours retain their familiar numbers on the 24-hour clock but don't need the bothersome A.M. tag. Six in the morning in 0600, 10:30 A.M. is 1030, and noon is 1200. Afternoon hours don't need to be marked with a P.M. but are numbered from 13 through 24. So 1:45 P.M. on the 12-hour clock becomes 1345, 7:05 in the evening is 1905, and so on throughout the day. Midnight can be expressed two equally correct ways, 2400 on November 15 being the same time as 0000 on November 16.

■■■■■■■■■■■■■■■■■■■■■■■■■■■■■■■■■■■■

interval from one noon to the next defines a solar day and tells us "how long." Our situation relative to the day's noon determines time epoch and tells us "when."

It sounds simple enough, but consider this: when it's noon in Kathmandu it's noon in neither Cleveland, Cracow, nor Kampala. Each of these cities shares the same solar day, yet each has its very own noon. Each has the same motion relative to the sun over the course of time, but none has the same position regarding the sun at any particular point in time. While the day may be a universally standardized quantity, noon is a local and relative event.

In a strictly solar sense noon occurs when the sun reaches its highest point in the sky. At this moment the sun is said to cross the zenith or to be on the meridian, an imaginary north-south arc across the sky that passes directly overhead. The hours before this solar climax are designated A.M. for *ante meridiem,* or before midday. The hours following it are designated P.M. for *post meridiem,* or after midday. Midnight occurs precisely halfway between one solar noon and the next. The rest of the points on the clock are uniformly staked out in relationship to these landmarks.

As the earth spins on its axis each spot on its face experiences one noon per rotation, or one per day. During any given 24 hours noon occurs everywhere once and only once. The motion of the earth from west to east results in a noon that sweeps across its face from east to west. This moving noon completes one circuit of the planet every 24 hours, moving steadily and relentlessly. It's always exactly solar noon somewhere. Some portion of the globe is always directly facing the sun. (See Figure IV.1.)

Until fairly recently this dynamic nature of time epoch was of no special concern. The Sumerians had their own noon as did the Babylonians, the Egyptians, the Greeks, and the Romans. Each was based on the overhead passage of the sun at their particular location. None really cared, if they even knew, that their noon was uniquely their own. Even within a given civilization the difference in time from one location to another went unnoticed. How was a Greek in Sparta, say, to know what time it was in Troy? Without telephones, radios, or any other form of instantaneous communication the only way to find out would have been to travel to Troy. And, once in Troy, the traveler would have had no way of knowing what time it was back in Sparta.

It wasn't until the development of accurate portable timepieces that the difference between noons in various places could even be documented. A merchant traveling from Paris to Rome in the seventeenth century could and probably would have been carrying a watch set to his native Paris time. In Rome his fellow traders would be functioning according to Roman solar time. When solar noon occurred in Rome, and the clocks struck twelve, the French watch would show about 11:20 A.M. When this watch finally got around to 12:00, when the sun passed overhead in Paris, Rome would already be some 40 minutes into the afternoon and the sun there a bit beyond the meridian.

This existence of different solar times presented no great inconvenience at first. The French merchant merely had to adjust his watch to Roman time when in Rome, and reset it to Parisian time upon his return home. While in Rome, Parisian time was of no practical concern, and vice versa. Throughout Europe, for hundreds of years, each town had a clock tower that chimed out the hours based on its own particular solar noon. Each was essentially an independent timekeeper, neither dependent upon nor answerable to any other. And each was as correct as the next.

The strictly local nature of time only began to be a nuisance with the advent of more rapid transportation and communication. The use of trains and the telegraph in particular gathered together the separate strands of civilization and wove them into a common fabric. It was then that the simultaneous existence of hundreds of time epochs became downright annoying rather than mildly amusing. The railroads found that all these different times resulted in some real headaches when making up their schedules. The one o'clock train from Philadelphia to Pittsburgh, for example, might need ten hours to make the trip, but wouldn't arrive at 11 o'clock, but at 10:40. The train from Pittsburgh to Philadelphia, meanwhile, might also leave at one o'clock and require ten hours of travel time, but not arrive until 11:20.

To cope with this mess each railroad adopted its own standard time that was consistent along all its routes. This time was generally set to solar time in

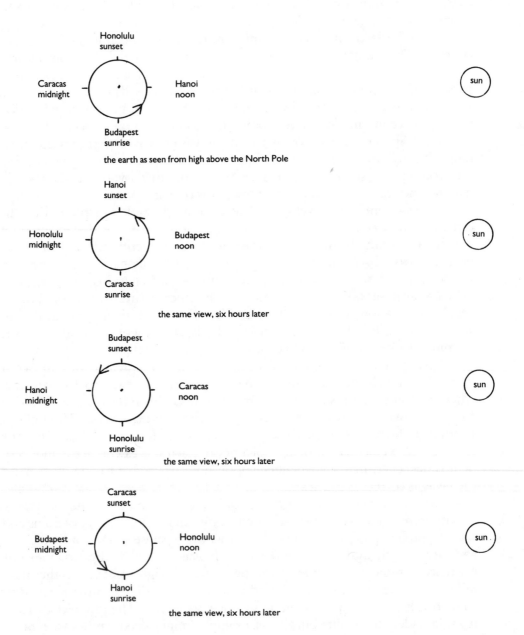

Figure IV.1 A spinning earth in a static day

the carrier's home city and ignored all other solar times. While this arrangement produced some solutions for individual railroads—the Pittsburgh-Philadelphia express now took ten hours by the clock in each direction—it didn't help the stations serving more than one line or the people trying to use them. Consider a commuter standing in New York's Grand Central Station at solar noon. She would be confronted by a whole bank of clocks, one for each connecting railroad. The clock for the B&O Railroad, set to Baltimore time, would read 11:50 A.M.; the clock for Penn Central, set to Philadelphia time, would read 11:55 A.M. Meanwhile, her wristwatch, set to her home in Boston, would show 12:12 P.M.

This situation was finally resolved in 1883 when the various railroads operating in the United States and Canada, meeting at the General Time Convention, decided to standardize time. They divided the North American continent into a series of time zones that closely resembled those in use today. The time everywhere within each zone was made the same, regardless of specific location or railway affiliation. The time difference between one zone and the next was set at exactly one hour.

It was an idea whose time had truly come. In the United States the plan was enacted into law as the Standard Time Act. In 1884, at the International Meridian Conference, an enlarged version of the plan, consisting of 24 time zones that girdled the entire globe, was approved in principle. It has since been gradually accepted and used. Nearly everywhere in the world Standard Time has replaced what is variously known as local apparent time or apparent solar time.

IV.15 | TIME ZONES

It's more than mere coincidence that there are 24 standard time zones, each one an hour different from its neighbors, as well as 24 hours in a day. For just as it's always solar noon somewhere, so it's also always one o'clock somewhere, two o'clock somewhere, and three o'clock somewhere else. At any given moment all the times of the day are being experienced simultaneously across the face of the globe. The sun is always shining, setting, or rising somewhere. Some portion of the earth is always being exposed to the sun and some portion is always turned away.

Because the planet spins on its axis, the portion of its surface being exposed to the sun is constantly changing. Each terrestrial location experiences its own particular solar noon as it in turn assumes a position directly facing the sun. In a strictly solar sense, there is a continual sweep of noon across the face of the globe that hits everywhere once every day. What standard time does is break this unbroken spectrum of solar time down into 24 one-hour chunks. Instead of an endless progression of solar noons there are just 24, occurring at one-hour intervals.

Each time zone experiences one noon per day as the earth rotates. Like horses on a carousel the zones remain permanently in step but forever behind or ahead of one another. Within each time zone the time of day is consistent. Within each time zone the time of day is one hour later than in the next zone to the west and one hour earlier than in the next zone to the east.

Each zone, in theory, includes 1/24th of the earth's surface and is shaped like the rind from a wedge of cantaloupe. In reality the zones are rather irregularly shaped so as to correspond as closely as possible with political and geographical boundaries and generally cause as little inconvenience as possible. In order to make them as functional as possible, though, they still approximate the geometrical ideal.

Time zones are organized along lines of longitude, imaginary divisions of the earth's surface that run from pole to pole along directly north-south routes. Each line of longitude is numbered according to its position. The 0° line of longitude runs from the North Pole directly south through Greenwich, England, and on to the South Pole. It's called the prime meridian and was established as such in 1884 by an international group of astronomers, cartographers, and timekeepers. To the east of Greenwich lines of longitude are numbered from zero to 180 and given the letter *E,* for east. To the west of Greenwich they're similarly numbered but given the letter *W,* for west. Longitude 180°E and 180°W are the same, and lie on the side of the globe directly opposite Greenwich. The 180° line of longitude and the prime meridian together form a complete circle around the earth and divide it into the Eastern and Western Hemispheres.

Because the entire globe consists of 360° of longitude—180° east plus 180° west—each time zone, on the average, consists of 1/24th of 360°, or 15° of longitude. Each zone is centered about its own standard meridian. Greenwich Standard Time, also called Universal Standard Time, is centered about the prime meridian and extends 7.5° to either side of it, more or less. In the United States, Eastern Standard Time is centered about longitude 75°W, Central Standard Time about 90°W, and so on.

The longitude of any location, when known, can be effectively used to at least estimate its time relationship to another location of known longitude. For instance Paris, France, at around 2°E and Ulan Bator, Mongolia, at 107°E lie 105° apart. Seven time zones, at 15° each, are also equal to 105°. It's reasonable to assume, then, that Paris and Ulan Bator are seven hours apart. The Mongolian capital, lying farther to the east, remains always seven hours ahead of its French counterpart. When it's noon in Paris it's 7 P.M. in Ulan Bator. Meanwhile, in Quebec, Canada, at 73°W longitude lying 75° or five time zones to the west of Paris, it's 7 A.M.

Because the earth rotates from west to east it's only longitude, or relative east-west position, and not latitude, or north-south position, that matters where time epoch is concerned. A traveler going east from Frankfurt to Kiev will have to reset her watch because she has changed her position with regard to the sun. She'll have to make similar adjustments if she travels west to Winnipeg. But, if she travels due north to Oslo or due south to Tunis she won't. These cities, lying on the same longitude as Frankfurt, share the same solar time and lie in the same time zone. It's east-west travel that results in the need to compensate for time differences when calling home. It's also east-west travel that results in the disruption of the biological sense of time. This disruption, called jet lag, doesn't occur during north-south travel.

Traveling eastward one moves in the same direction as the earth rotates and thus accelerates the passage of the day. For each 15° of longitude traversed an hour is lost to the sun. Traveling westward the reverse is true. The traveler's motion counteracts the rotation of the earth and the passage of the day is retarded. Indeed, if one were to travel in a westerly direction at a pace of 15° of longitude per hour one would exactly offset the easterly rotation of the globe. One could, theoretically at least, remain forever positioned toward the sun and experience a perpetual solar noon. With the advent of jet travel this sort of situation is perfectly feasible, at least temporarily. It's also possible to travel at a rate of more than 15° of longitude per hour. In this case one can actually move faster than the earth's rotation, gain on the sun, and reverse the progress of the day.

There is, unfortunately, a limit to how much one can cheat the clock in this manner. A traveler flying west at a rate of 15° of latitude per hour can't really expect to remain forever suspended in time. A traveler circumnavigating the globe at a rate in excess of 15° per hour can hardly expect to leave Greenwich at noon and return there earlier the same day. In each instance the globe will be spinning beneath him and time marching onward during the course of his journey. In each case the time he gains by setting his watch backward when he crosses successive time zones will be offset as he sets it forward when he crosses the International Date Line.

The International Date Line, which lies more or less at 180° longitude, is where one day ends and another begins. (See Figure IV.2.) As one crosses the date line going in a westerly direction one not only sets one's watch one hour earlier, as is the case with other time zone boundaries, but one day later as well. This net result of 23 hours forward exactly offsets the combined 23 hours backward of the other 23 time-zone changes. Going in an easterly direction the process is completely reversed. A traveler loses an hour every 15° until he crosses the date line, at which point he gains 23. Any circumnavigation consists of 23 one-hour leaps in one direction and a single 23-hour leap in the other.

Generally speaking, the date on the western side of the International Date Line is one day later than it is on the eastern side. Only when midnight occurs along the line (when it's high noon in Greenwich) does the entire planet reside in the same little box on the calendar. At all other times some of us are finishing up one day while the rest of us have already started on the next.

In a technical sense, of course, the day never really begins or ends but remains steadfastly in place as the earth spins within it. Time zones, the International Date Line, and all the attendant problems of dealing with time epoch stem ultimately from our need to break up the continuous bath of sunlight upon the continuously rotating earth into a series of distinct units

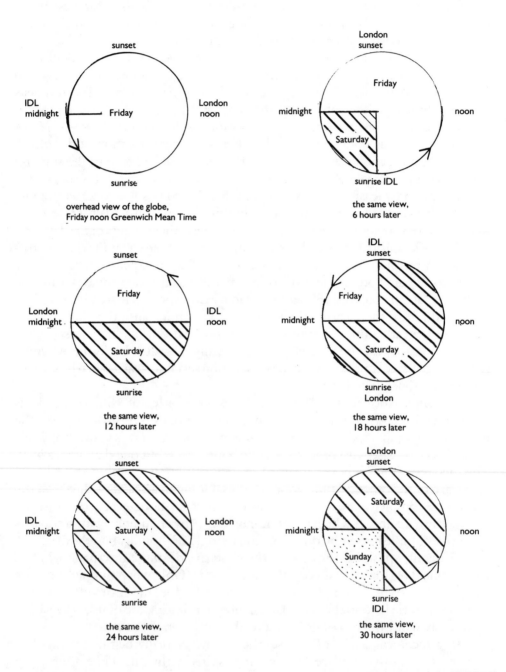

Figure IV.2   The operation of the International Date Line (IDL)

called days. They all stem from our need to impose artificial structure on natural, free-flowing time.

IV.16 THE END OF TIME

Time interval and time epoch have both been defined, calculated, and legislated with enough precision and vigor to satisfy the fussiest time fanatic. But, to be realistic, all we've done is create mechanisms that allow the rather limited human intellect to cope with something that lies far beyond its capabilities. Time exists apart from any of our tools and inventions. It existed long before there were any human minds to contemplate it and it will continue to exist long after they're gone.

Indeed, the origins of time have been traced back billions and billions of years to an era that predates our race, our biological ancestors, our planet, our sun, and our galaxy. Some 17 billion years ago, it's thought, the universe as we know it was born in a colossal explosion that created matter, energy, and space, and hence time. (See Unit IV.5.) What existed before this remotest imaginable event can't even be guessed at, for there's absolutely no evidence of it. The end of time, it's suspected, will also coincide with the end of reality as we know it. When time ends, so will motion, change, and progress. Is this likely to happen? If so, when?

To the proponents of a steady-state universe, of course, the final apocalypse will never happen. The universe will continue to expand forever and new matter will continue to be generated to fill it. This is an appealing theory emotionally. It provides unlimited time and opportunity, avoids any inescapable finality, and operates on a basis of perpetual renewal and growth. Such a future fulfills the human need for hope and immortality of some kind. Unfortunately, it may be just these sorts of human needs that resulted in the theory of a steady-state universe in the first place. Accumulating physical evidence makes its existence increasingly unlikely.

Most physicists are now fairly confident that the universe we inhabit is a Big Bang universe. A Big Bang universe, unlike its steady-state alternative, is undergoing no more creation. A Big Bang universe contains all the matter and energy it will ever have. As it expands its given quota of these commodities simply becomes more and more diluted over time and space.

A universe that continually expands and dilutes itself will eventually die what's called an entropy death. Entropy, remember, is a state of disorder or randomness, and many natural processes proceed only in directions that increase entropy. None spontaneously proceed in directions that reduce entropy. (See Unit IV.2.) Entropy death would occur when entropy could be increased no further and no more change could take place. Without the change that makes it meaningful, time would cease to exist.

The arrival of an entropy doomsday, although ridiculously remote, has been given a date. Time will cease when change ceases, and change will cease

When and Where

The development of accurate timepieces was motivated to a great extent by the needs of sailors. Ever since the year 1100 or so seafarers had been able to determine their direction by the use of a compass or its crude forerunners. They had been plotting their north-south location on the globe, what's now known as latitude, for almost as long by the use of a variety of instruments like the sextant, which measured the elevation above the horizon of the stars.

But 200 years after the first circumnavigation of the globe there was still no way for a ship at sea to reliably determine its longitude. East-west positions were still being charted by the ancient practice called dead reckoning. This was really nothing more than educated guesswork that calculated distance by multiplying estimated speed by the time traveled.

It had been known since 1530 or so that the solution to the problem was the interrelationship of longitude and local, or solar, time. If a ship experiencing solar noon knew what time it was in Greenwich at that same moment it could then calculate its east-west position relative to the prime meridian. If noon on board occurs when it's 2:00 P.M. in Greenwich, for example, then the ship must lie at 30°W longitude. Determining longitude this way, of course, requires a chronometer capable of keeping an on-board record of Greenwich Mean Time.

In 1714, after a rash of maritime disasters resulting from navigation by dead reckoning, the British government established the Board of Longitude and offered a sizable cash reward to anyone building such a device. The reward was finally claimed in 1761 by John Harrison (1693–1776), a carpenter. He designed a number of instruments, the best of which kept time to within three seconds a day over the course of a six-week sea journey, a performance then unmatched by any land-based clock.

—Can you explain how the sextant measures latitude? Can sightings of the sun as well as the stars be used this way?

—Early attempts to keep track of what time it was in Greenwich relied on observing the moons of Jupiter through a telescope. How did this work? What were its drawbacks?

—Why were the clocks of Harrison's day so unreliable at sea?

—Suppose you know the exact distance you have sailed due west from the prime meridian. Can you calculate your longitude? What if you are unsure of your latitude?

■■■■■■■■■■■■■■■■■■■■■■■■■■■■■■■■■■

The Big Crunch

The ongoing expansion of the universe is thought to be the result of a massive explosion called the Big Bang. Any contraction and eventual collapse of the universe will have to rely on gravity, the mutual attraction of all matter. Overcoming the effects of the Big Bang will require a critical amount of gravitational force and sufficient matter to generate it. According to current calculations, the required matter far exceeds that which we know exists and the so-called Big Crunch will never happen.

Proponents of an oscillating universe aren't overly concerned, however. New, more powerful telescopes continue to reveal the presence of more and more stars. Other techniques have also revealed matter that will never be seen with any telescope. Radio astronomy, for example, can detect the presence of matter too cool to emit light waves but warm enough to emit lower-frequency radiation. And, via indirect means, astronomers are also attempting to verify the presence of black holes—bodies so massive that their gravitational fields prevent the escape of any radiation.

■■■■■■■■■■■■■■■■■■■■■■■■■■■■■■■■■■

when the slowest-known natural process has run its course. This slowest process is evidently the decay of a rare radioactive isotope of lead, lead-204. This radioactive disintegration has a half-life of about 14 billion billion years. It will take an estimated one thousand halvings to totally eliminate all lead-204 from the universe. One thousand halvings will take one thousand half-lives, or 14 trillion billion years. By this time all the planets will have long ago been slowed to a standstill by friction with cosmic dust and will have crashed into their respective suns. By this time all the stars will have burned out. By this time all the heat in the universe will have flowed from hot spots to cool ones to create an all-pervasive, uniform chill. Based on a life span of 14 trillion billion years the universe now is a newborn infant.

But entropy death, inevitable in a continually expanding universe, may never arrive. The universe may not expand forever. The proponents of an oscillating universe predict that expansion, under way now for some 17 billion years or so, will proceed for only another 23 billion years. Then, when the forces of gravitational attraction overcome the explosive forces of the Big Bang, a 40-billion-year period of contraction will ensue. If we do, indeed, reside in an oscillating universe, it may well be only one of a whole series of universes that are born, expand, contract, and collapse at 80-billion-year intervals. Each Big Bang may, in fact, be nothing more than a single heartbeat in the life of the universe—a single tick of the cosmic clock.

"Man is the measure of all things."
—Protagoras, c.485–c.410 B.C.

V.1 A CONFUSING PREDICAMENT

Lavoisier stalking the elements, Curie probing the source of nuclear energy, Copernicus charting the heavens, and the ancient astronomers chronicling the seasons all had something in common. Certainly, all were examining the physical world in search of answers to practical questions—all were trying to add security, comfort, power, or predictability to human existence. But beyond that, all were working on different aspects of a single, much more grandiose project. All were trying to figure out just how the world works.

Each and every one of us is, to some extent anyway, trying to do the same thing. We may not spend our days toiling away in research labs or manipulating complicated equations, but in our own way we're all constantly gathering bits of information about the world we inhabit and arranging that information into patterns. We are, in short, trying to make some sense out of the complex and confusing existence into which we've been so unpreparedly thrust.

This urge to explore and understand is nothing less than human nature. It's what sets us apart from the beasts. Granted, this unquenchable curiosity of ours may one day prove to be our undoing, but it has, up to this point anyway, helped our species flourish while others have perished.

The first investigations of reality were undertaken with energy, urgency, and good intentions, but in the complete absence of higher thought or method. There were no hypotheses proposed, no experiments conducted, and no theories formulated. Even the most fundamental principle, that of cause and effect, was unrecognized. Still, there was progress. The relentless use of trial and error over thousands of years, together with sporadic moments of coincidence and serendipity, gave us fire and the wheel—innovations whose importance remains unsurpassed. With fire our ancestors were able to ward off the advances of predators and ensure their survival. With the wheel and other simple tools they were able to multiply their strength and make survival a bit more comfortable.

Later, the domestication of animals lent the wheel even greater utility. A domestic livestock supply eliminated humans' dependency upon the vagaries of wild-animal migrations and bolstered their ability to survive disease, drought, and other hardships. Finally, with the domestication of certain plants—the grains in particular—human populations were able to attain a certain degree of self-sufficiency, forsake nomadic wanderings for the stability of settlements, and develop and accumulate the trappings of that wondrous thing called civilization.

The establishment of permanent camps and their subsequent growth into villages, towns, and cities provided previously unknown concentrations of people. This, in turn, made possible the division of labor, the development of specialized skills, and the exchange of ideas and materials that fueled progress and raised the standard of living. The use of fire became increasingly sophisticated. It was used in kilns to bake pottery, in furnaces to make glass, and in smelters to forge metals. The wheel was also improved upon to make the wheel and axle, the pottery wheel, the pulley, the drill, and the screw. A budding technology began to blossom.

Another unparalleled invention, language—first spoken and then written—ensured that any progress made was preserved and passed along to future generations and even other cultures. Fire, once discovered, never went out. The wheel, once invented, never stopped rolling. With the efforts of their ancestors secured, later generations were free to concentrate on new problems and find new solutions. This cumulative nature of knowledge made the relationship between man and reality an ever-changing one. Slowly, ever so slowly, the balance of power between nature and its human captives began to shift.

Nature, however, remained firmly in charge, incomprehensible and for the most part uncontrollable. Because of its overwhelming power, its inexplicable behavior, and its potential fury it was a source of awe, dismay, and fear. It was dangerous. It was an adversary. This adversary was conceptualized in terms of the only other thing humankind was familiar with that possessed such a mercurial temperament—itself. Thunder and lightning, the wind and the rain, the sun, the moon, the stars, the sea, fire, the dark of night, the seasons, and many other aspects of reality were assigned identities that were humanly whimsical, passionate, and erratic.

Because of their immense power, however, these natural personalities assumed superhuman proportions in the minds of their creators. They became gods. Mortal humanity, living in their midst, was at the mercy of their alternating kindnesses and cruelties. To appease them we sang songs, chanted prayers, held elaborate ceremonies, offered sacrifices, and built grand temples. The first relationship between humanity and the rest of the universe, then, was a religious one, and foremost among its practitioners were priests.

V.2 NATURAL PHILOSOPHY

As the years piled up so did the amount of information collected, catalogued, cared for, and studied. It became increasingly apparent that reality wasn't a random series of events, but proceeded according to a design of some sort. The regularity of nature was most easily recognized in the heavens. The sun's daily journey across the sky, bringing day then night, was the very epitome of regularity. The seasonal ascent and descent of the sun's overhead arc, accompanied by longer, then shorter days, was similarly predictable. The wandering stars—the planets—also moved along pathways that, although complicated and unrelated to the sky in general, at least formed some sort of pattern. Even solar and lunar eclipses seemed to adhere to a fairly rigid, if not completely legible, schedule.

The priests, in order to enhance their prestige and maintain their privileged positions, made it their business to learn as much as they could about these natural rhythms. They became fledgling astronomers. Temples became observatories. Religion, in some aspects at least, became somewhat scientific. Reality was still conceptualized in superhuman terms, but these all-powerful beings were starting to seem a little less emotional and a bit more rational. The gods were most certainly mysterious and wondrous, but they weren't necessarily crazy.

Such were the prevailing conditions and attitudes when, about 600 B.C. or so, across the Grecian peninsula and islands, various city-states banded together to form a civilization the likes of which had never been seen before and which, in some regards, has never been seen since. Operating under the novel and radical tenet that knowledge was good in and of itself, the Greeks sought to amass as much of it as they could and distribute it as widely as possible. The aristocracy of the priesthood was democratized, its monopoly on power made a public trust. In its place there arose a sizable, diverse, well-educated, idealistic, intelligent, and extremely inquisitive population of thinkers, orators, writers, poets, dramatists, mathematicians, engineers, musicians, linguists, architects, and artists. Collectively they were known as philosophers, or lovers of wisdom.

Their accomplishments were spectacular. Thales, combining historical data with powerful insight, correctly explained the mechanism of eclipses. Using objective observations and disciplined logic he demonstrated that the world was round. With careful measurement and sound geometric principles Eratosthenes calculated the circumference of the globe. Employing similar techniques Hipparchus computed the distance to the moon. (See Unit III.2.) In every instance it was human intellect, not divine whimsy, that most convincingly explained the operation of the universe.

The Greeks were understandably impressed with their accomplishments. As one problem after another fell before the onslaught of logic and mathematics they began to assume that with enough time and thought the entirety of human events and experiences could be reduced to a few formulas. And, indeed, in the areas of ethics, law, geometry, logic, trigonometry, architecture, music, and military strategy they often achieved this lofty goal.

But in the area of natural philosophy, what contained the rudiments of what we now call science, the results were less favorable. Combustion, motion, and the great array of objects and substances that populated the material world behaved in a manner that was far less straightforward than that of the heavens. And while the application of the principles of geometry or ethics could be restricted to very specific and tightly controlled ideal conditions, the natural philosophers were forced to operate out in the great big messy and very real world. Regardless of the eloquence of their arguments or the soundness of their logic, they could never escape the realities by which their theories were judged.

Unable to satisfactorily explain the operation of the world around them, the natural philosophers grew increasingly frustrated. They began to ignore more and more of the annoying details and to focus more and more on grandiose abstract concepts. Eventually natural philosophy deteriorated into little more than an academic exercise. This abandonment of the real for the ideal was justified by claiming that everyday objects and events were of only limited use and not all that interesting. What truly mattered was the exercise of the mind in realms of unlimited possibilities and potential.

The undisputed champion of this sport was Aristotle. His fanciful universe consisted of two distinct regions: the terrestrial and the celestial. The terrestrial region, in which the body was regrettably trapped, was a hopelessly defective place that couldn't be properly described in mathematical or logical terms. The celestial region on the other hand, physically off-limits but fair game for intellectual exploration, was an elegant and philosophically proper locale that obeyed higher truths.

The lower terrestrial realm consisted of four types of common matter—water, fire, earth, and air—inextricably jumbled together. The upper celestial realm was 100 percent pure aether. Earthly objects were of random shapes and sizes and were either static or in awkward states of irregular and temporary motion. Heavenly objects were unblemished spheres that moved forever in geometrically precise circles. (See Units I.2, II.4, and III.3.)

All objects, regardless of the realm they occupied, were driven to do what they did according to the dictates of their inner appetites. Because of its content of levity it was the appetite of smoke to rise. Because of its content of gravity it was the natural tendency of a grape to fall. An apple, possessing more gravity than a grape, had a greater appetite and so fell faster and with a greater force. And the planets circled the earth simply because it was their nature to do so.

Although full of contradictions and essentially worthless from a practical point of view, this Aristotelian version of reality was nonetheless an improvement over its forerunners. The world was no longer ruled by supernatural powers, but by natural appetites that, although a long way from being understood, were at least consistent and followed rules of some kind. The logic of philosophy had replaced the emotion of religion.

V.3 | THE BIRTH OF SCIENCE

Aristotle's highly stylized version of the universe, despite its obvious shortcomings, enjoyed a popularity that spanned some 1,500 years. At first it was Roman indifference and then medieval neglect that sustained it. Later, it was the Catholic Church that validated the views of a man long dead who had declared that an apple fell to earth faster than a grape without ever having taken the time to substantiate his claim.

The Church needed some sort of explanation for worldly events and found the reality of Aristotle quite convenient for several reasons. First of all, Aristotle's work was easy to incorporate into Church policy because it already existed as a self-contained and comprehensive body of thought. Secondly, Aristotle was well known and had the sort of respect the Church wanted. Finally, the world designed by Aristotle was fairly compatible with Church doctrine, having been rid of spirits, mysticism, and magic.

There were, of course, a few modifications to be made. Aristotle's world was a collection of objects motivated by their own appetites. It was an essentially valueless place that ran along pretty much of its own accord, with no specific purpose other than to entertain philosophers. The Church placed the one true God in charge of this universe and placed it in a broader religious context. No longer a challenging intellectual puzzle, it was now a titanic struggle of good against evil with the fate of humankind hanging in the balance.

Once embedded within the creed of the Church the ideas of Aristotle were virtually impossible to dislodge. Indeed, if *Homo sapiens* consisted of nothing but average men and average women producing average sons and average daughters we might still be living in the Dark Ages. But it doesn't and we don't. For better or worse, due to the quirks of genetics and circumstance, we find among us certain rare individuals able to rise above the treetops and see the surrounding forest. The overthrow of the Aristotelian version of reality and the subsequent birth of modern science was accomplished by a rather small collection of these visionaries operating over a relatively brief span of time. Greatly simplified for the purposes of illustration, the process can be faithfully represented by three men whose work encompassed a span of about 150 years.

The first of these men was Copernicus, an unlikely, even reluctant revolutionary. After struggling mightily to justify the scheme invented by Aristotle and approved by the Church, Copernicus was compelled by his own doubts and genius to conceive the inconceivable. Against all conventional wisdom and common sense he proposed a moving earth that orbited a stationary sun. His publication of *Revolutions of the Heavenly Bodies* in 1543 was a revolutionary event in every sense of the word. (See Unit III.4.) By merely suggesting that maybe everything wasn't the way it appeared to be or was supposed to be Copernicus first applied the chisel of doubt to the carefully maintained facade of traditional beliefs.

Then Galileo took up the chisel and almost gleefully shattered the solid but brittle status quo. What had been doubt and speculation with Copernicus was disbelief and defiance with Galileo. Using telescopes, thermometers, and pendulum timers of his own construction he proved that all bodies fell at the same rate regardless of their weight. (See Unit II.5.) He demonstrated that air had weight. (See Unit I.24.) With his own eyes he saw the moons of Jupiter and could say with certainty that not everything revolved around the earth. (See Unit III.6.) It was Galileo who first practiced the art of measurement, successfully interpreted natural events in the language of mathematics, and refined the experimental method. What had first been the jurisdiction of priests and then philosophers was, starting with Galileo, the business of scientists.

The attitudes and procedures developed by Galileo enabled Newton to vault science from its infancy into maturity with breathtaking speed. His *Mathematical Principles of Natural Philosophy,* published in 1687, defined the fundamental concepts of force, motion, space, time, and mass in terms that are still used. His mathematical relationships among these basic entities described a physical world that equalled or surpassed that of Aristotle in terms of completeness and coherence. But Newton's model corresponded to daily reality whereas Aristotle's didn't. Newton explained things not as they ought to be or as he would have liked them to be, but as they really were.

The contents of the *Principia* are known now as Newtonian, or classical, physics. If judged in terms of impact, classical physics, along with its offspring, must surely be considered one of the most powerful networks of ideas ever formulated by the human mind. It's classical physics that has transformed the candle-lit, horse-drawn, homespun world of 1686 into the fluorescent, fuel-injected, polyester world of today. Compact discs, freeze-dried coffee, fax machines, air-conditioning, permanent-press shirts, ICBM's, and artificial sweeteners are all products of a technology that operates according to the principles of Newtonian physics.

With this sort of performance to recommend it, it's easy to see how classical physics or, more broadly, science in general, could be seen as the ultimate solution to the puzzle of reality. And so it was during the eighteenth and nineteenth centuries when it enjoyed a series of seemingly endless successes and virtually no failures. At its zenith the heady advance of science led to the popularity of a school of thought known as determinism.

Newton had shown how great chunks of matter, the planets, subjected to the physical forces of inertia and gravity, moved along utterly predictable and minutely describable paths. (See Unit III.9.) He had also shown how smaller chunks of earthbound matter responded to pushes and pulls in a similarly straightforward manner. (See Unit II.6.) There seemed to be no logical limit to the scale on which this principle could be applied. If, as theorized and later demonstrated, matter consisted of tiny little chunks called atoms, then each of these chunks, too, must behave according to the universal laws of pushes and pulls known formally as mechanics.

The world, said the determinists, consisted of a finite number of atoms each with a certain mass, location, speed, and direction. Each of these atoms behaved in accordance with Newton's laws of motion, they maintained, and was therefore totally predictable. Seeing as how it was the interaction of these particles that determined the course of the universe, the future was an inevitable consequence of the present. Free will, choice, and chance were imaginary. Everything happened inevitably as the result of matter obeying the laws of motion. Given enough data and a mind capable of performing the required calculations, the determinists were prepared to know everything.

■ ■

Is There a God?

Throughout history the actions of a god or gods have been routinely called upon to explain everything not otherwise understood. An accumulation of scientific breakthroughs has accordingly resulted in an ever-diminishing role for deities. Indeed, the determinists believed that God was needed for nothing beyond creating the universe and then setting it into motion according to the laws discovered by Newton. Everything else, they contended, proceeded inevitably from that point.

The latest developments in theoretical physics indicate that the universe may be unbounded in time—that is, it may never have had a beginning—and that even the laws it obeys may be inescapable consequences of its mere existence. Are we ready, then, to dispense of God altogether?

Scientists generally answer this question with emphatic denials. Even if we eventually succeed in explaining every aspect of how the universe works, they say, we'll never be able to determine why it exists in the first place. For those who ask this ultimate question, God (or god, or gods) will always be needed to provide an answer.

■ ■

V.4 | A FATAL FLAW

The determinists, of course, were dead wrong. Granted, classical physics was a far superior means of explaining the physical world than the philosophical conjectures of Aristotle, but it was by no means a perfect or final solution. As we have since discovered, Newton's scheme contained a fatal flaw or two that eventually proved to be its undoing. Although no one could say for sure just what the shortcomings were at first, there were those who suspected something was amiss from the very moment Newton unveiled his radical ideas.

Momentous revisions of thought and attitude, like those demanded by Newton, don't come easily or quickly. They must inevitably overcome the natural resistance to change and the fear of the unknown. They must survive close scrutiny and be judged harmless. Newton's explanation of the universe was greeted not only with accolades, but with skepticism as well. How could the gravity of the earth reach all the way to the moon? How did it get there? How could lumps of inert matter generate such enormous amounts of force? And if everything really did attract everything else, why didn't the stars all come crashing down in a giant heap?

Newton, aware of both the proper use and the limitations of science, refrained from even asking these sorts of questions, much less trying to answer them. "I do not deal in metaphysical speculation," he declared. By confining himself to the "what, where, and when" rather than the "how and why" he accomplished momentous things—he described simultaneously the fall of an apple and the operation of the solar system. His detractors, though, still stuck in the natural philosophy mind-set of the Greeks, couldn't resign themselves to such a limited yet powerful attitude. Unimpressed by a few clever formulas, they wanted to know how gravity worked.

In order to make gravity emotionally acceptable, then, there needed to be a physical medium of some sort through which it operated. In order for the stars to be suspended in the sky there had to be a fluid of some sort in which they floated. There had to be some sort of stuff that allowed all the otherwise unconnected pieces of the universe to function interactively. The required substance would have to be weightless, crystal clear, and universal. Luckily, such a substance already existed.

Aristotle's universe, remember, was of two quite distinct parts. One part was messy and unmanageable, the other part sublime and orderly. The familiar but flawed portion was composed of earth, water, air, and fire. In the exotic reaches of outer space, meanwhile, it was pure, crystalline aether that carried the sun and the moon across the sky, that held the stars in place, and through which the planets wandered. Here, most certainly, was the means by which gravity operated. And so aether was resurrected, modernized as "ether," and injected into Newton's physics to make it more philosophically pleasing.

Newton himself resisted the introduction of the revised and respelled ether into his universe. He saw no physical evidence of its existence, only an emotional need that dated back to Aristotle and his appetites. Still, he had to admit that it was difficult to imagine "that inanimate brute matter should, without the mediation of something else . . . operate upon and effect other matter without mutual contact." Although undocumented, ether would be convenient. What would be the harm?

As new discoveries about the physical world continued to mount, the case for ether grew stronger and stronger. Newton, having enjoyed great success describing the universe in terms of chunks of matter and physical motion, had assumed that light, too, consisted of very tiny bits of stuff traveling very rapidly from a source to a destination. The subsequent investigation of light, however, suggested that it behaved more like a beam of pure energy than like a spray of particles. Light, like gravity, was capable of leaping across enormous distances without any apparent means of conveyance. Surely there had to be something, an undetected fluid of some sort, that conducted light from point A to point B. Magnetism, too, demanded an agent that accounted for action at a distance. So did radiant heat. All needed something to connect cause and effect.

If the mechanistic system of Newton was to be believed, if reality operated like a giant clockwork of interconnected parts, then there had to be a universal connector of some sort that explained all these otherwise inexplicable events. There had to be ether.

V.5 THE DEFINITION OF MOTION

There was yet another troubling aspect of the Newtonian explanation of the physical world that neatly disappeared with the supposition of ether. The whole of mechanistic, classical physics, including its extremist form of determinism, was based on the premise that reality consisted of nothing but matter in motion. In order to give this scheme a solid foundation it was necessary to have precise, accurate, ironclad definitions of the two basic agents involved: matter and motion.

The definition of matter presented no immediate problem. It had mass, occupied space, and generated and was influenced by gravity. The satisfactory definition of motion proved to be not quite so simple. It was, in fact, an extremely complicated issue whose roots stretched back to Copernicus.

Since antiquity motion had always been easily understood in relationship to the stationary frame of reference in which it occurred. This frame of reference was rather obviously assumed to be the solid earth upon which all events and their observers resided. But when Copernicus proclaimed that the earth orbited the sun he gave motion to what had previously been motionless. He changed the long-standing and unquestioned reference point for motion from the earth beneath us to a distant sun.

The complications arising from this change of reference point went unnoticed at first, for they were but minor details compared to the other issues Copernicus raised. Initially the emotional, psychological, and religious aspects of rearranging the universe were of paramount concern. The fact that we were no longer the focal point of the universe is what tormented Copernicus, outraged his attackers, and fueled the controversy that resulted in Galileo's conviction and house arrest by the Church. Certainly the physical ramifications of the fact that we and our apparently stationary planet might be whizzing around the sun at a seemingly impossible speed were hotly debated. But, for the most part, physical issues and evidence were used only as secondary arguments to either support or refute positions taken based on the psychological aspects of the revolution.

As it turns out, though, it's the physical and not the emotional consequences of putting the earth in motion that have had the most impact on the evolution of science and our concept of reality. These complications were first hinted at by Galileo when he took up Copernicus's cause and attempted to defend it with physical rather than emotional arguments. Once the physical objections had been met, he felt, psychological ones would weaken. So he had to convince an understandably skeptical public that the earth moved. It wasn't easy. It was, in fact, impossible. Galileo didn't know about the aberration of starlight or stellar parallax. (See Unit III.10.) He predated Foucault's pendulum. (See Unit III.11.) Given the knowledge and technology available to him Galileo was theoretically limited to proving only that the earth *could* move, not that it necessarily did. To this end he presented the following illustration:

> I invite you to enter with me the cabin of a large ship. In it are gnats, ball players and a bowl of goldfish into which drops of water are falling from above. Whether the ship lies at anchor or moves uniformly through the water, the gnats will fly with the same ease from wall to wall; the balls will be hit with the same force in all directions; the fish will swim undisturbed; the drops will fall vertically upon the same spot. No one can guess, by observing these processes, whether the vessel is at rest or moving—no more than we can say this about the earth.

The point was well made. From events aboard it is impossible to tell whether the ship is at anchor or in uniform motion across a calm sea. Similarly, from the surface of the earth, it's impossible to tell whether the planet resides firmly in place or glides smoothly around the sun. The ship, indeed, could be under sail. The earth, indeed, could be in solar orbit.

While it was Galileo's intent to disarm his adversaries with this argument, he unwittingly raised a point that would eventually prove to be the undoing of the classical physics that he himself inspired. Yes, it was impossible to tell whether or not the ship was under sail and, yes, it was impossible to tell whether or not the earth was in orbit. But, if no one could say for sure that the ship or the earth were stationary, then no one could say for sure that they weren't, either.

The ship, if it did move, moved only in relationship to the sea upon which it sailed. The earth, if it did move, moved only in relationship to the sun around which it orbited. Ignoring the sea, the ship couldn't properly be said to move or lie at rest. Ignoring the sun, the earth could just as easily be at rest as in motion. Motion, it seemed, was only meaningful in regard to a stationary reference point from which to observe and measure it.

But what was the proper reference point? At first it had always been the earth. Then it was the sun. But later it was found out that the sun itself was swirling about the nucleus of the Milky Way and that even the Milky Way was cartwheeling through intergalactic space. Everything, it seemed, was in motion compared to everything else. There was nothing that stood perfectly still. There was no ultimate reference point for motion. The definition of motion was elusive if not downright impossible. One of the two pillars of classical physics had nothing to support it.

Newton had sidestepped this issue by using the vacuum of space as a motionless sea through which everything else swam. But space in this sense was only a concept and not a material thing. In order to be at rest, or to possess any other physical characteristic, space had to be a physical reality, not an imagined convenience. The only way to rescue Newton was to give some sort of substance to his mathematical space. Ether, besides being used to conduct the forces of heat, light, magnetism, and gravity, seemed ideally suited to this purpose as well. Space wasn't empty, but full of ether just as Aristotle had said. It was motionless ether against which all motion could be measured and from which it took meaning.

V.6 JUNKING CLASSICAL PHYSICS

The existence of ether now seemed all but assured. The validity of classical physics depended upon it. All that was needed was to find it.

Not an easy task. Ether had no noticeable effects on the behavior of light other than to transmit it. It was completely invisible. Ether in no way retarded or disturbed the motion of the heavenly bodies that circulated through it. It was intangible and frictionless. It was also tasteless, odorless, weightless, and motionless. It was everywhere present but nowhere evident. It was next to nothing, yet in order to satisfy all the demands made of it, it had to be both thousands of times more fluid than air and thousands of times more resilient than steel. Strange stuff, to be sure. But possible, of course. Not only possible, but necessary. Faced with the dilemma of either accepting this weird fluid or junking classical physics, scientists overwhelmingly chose ether.

The search for ether was a long and diligent one. It reached its climax in 1887. In that year two physicists, Albert A. Michelson (1852–1931) and Edward W. Morley (1838–1923), conducted a series of experiments designed to determine the reality or fiction of ether once and for all. If the earth moved through a sea of ether, they reasoned, then it must, like a motorcyclist on a still night, create its own "ether wind."

If a motocyclist tosses an object forward as he rides, the object's speed relative to the road will be equal to the speed of the motorcycle plus the speed of the throw. Likewise a rearward-tossed object will have a speed equal to that of the cycle minus that of the throw. Michelson and Morley expected light rays originating from earth to behave similarly. Light launched forward, so to speak, from the earth should have a speed equal to the speed of light in a vacuum plus the speed of the earth through the ether. Light launched in the other direction should have a speed equal to that of light in a vacuum minus the speed of the earth.

To test their hypothesis Michelson and Morley painstakingly and repeatedly calibrated the speed of light into, with, and across the supposed ether wind. The results never varied. To their utter amazement every beam of light traveled at exactly the same speed. There was no difference of velocities. There evidently was no ether wind.

A belief in the existence of ether now required the acceptance of two mutually exclusive absurdities. The first was that ether, although real, was completely undetectable. The second was that light traveled through ether at the same velocity regardless of the speed of its source. The first of these absurdities was needed to salvage classical physics. The second violated the laws upon which classical physics was based.

There seemed no way to reconcile the validity of Newton's laws, the existence of ether, and the results of the Morley-Michelson experiments. There were plenty of attempts, though—things like the Fresnel drag coefficient and the Fitzgerald-Lorenz contraction effect. But none worked. Each proposed solution was more complicated and more absurd than the problem it tried to resolve. Enter Albert Einstein (1879–1955).

Einstein took the bold step no one else had dared to take. Look, he said, if light appears to have a constant velocity regardless of its source then let's just say it does and have it over with. If there appears to be no such thing as ether then let's just say there's no such thing as ether and be done with it. If there is no such thing as absolute motion or absolute rest let's just accept it as a fact. And if by accepting all these things we have to abandon classical physics, then let's just do so and move on to another, more faithful model of reality. The year was 1905.

V.7 | THE THEORY OF RELATIVITY

Einstein, then a shy, 26-year-old clerk working in the Swiss patent office and totally unknown to the scientific community, was able to take the step no one else dared take because he knew what no one else knew. Only Einstein was able to discard Newtonian physics because only he had something with which to replace it. Only Einstein, working in his spare time without credentials and using nothing more than paper and pencil, had the theory of relativity. (Einstein's demolition of classical physics was actually a two-step process. He published what is known as the special theory of relativity in 1905. Eleven years later he unveiled an expanded version of his ideas known as the general theory of relativity.)

This theory, which remains essentially intact as of this writing, differs from classical physics in several fundamental and important ways. To begin with it assumes, without trying to explain how or prove why, that light always travels through a vacuum at the same measured speed, usually designated as c. Light beams originating from earth, the sun, a distant star streaking toward the earth, or a distant star streaking away from the earth all travel at precisely c. So light, along with all its electromagnetic radiation cousins, behaves differently from any material object. This is quite possible, for light isn't a material object. Light is just light.

Secondly, the theory of relativity neither supposes nor requires the presence of ether to fill space. Ether isn't needed because space, according to Einstein, is a vibrant, active force field with properties of its own that extend beyond the mechanistic concepts of matter and motion. Space is neither the passive mathematical emptiness supposed by Newton nor the sea of ether supposed by his successors, but a very real thing capable of transmitting light, gravity, radiant heat, and magnetism.

■■■■■■■■■■■■■■■■■■■■■■■■■■■■■■■■■■■

A Celebrity Scientist

Almost against his will, Einstein was catapulted into a position of international recognition by his theory of relativity. In 1921 he was awarded the Nobel Prize in physics, and shortly thereafter was recruited by the League of Nations as a spokesman. Humble, benevolent, and an articulate and dedicated advocate of world government, Einstein was perfectly suited for the position.

In 1939 it was Einstein who wrote to President Franklin Roosevelt informing him of the potential of nuclear energy. Einstein's reputation as a great thinker and a pacifist helped persuade the president to launch the effort that culminated in the development of the atomic bomb and the conclusion of World War II. (See Unit II.31.)

Although born of Jewish parents, a supporter of Zionism, and deeply religious, Einstein didn't associate with any orthodox faith. He was, nonetheless, offered the presidency of Israel in 1952. Perhaps mindful of a dismal performance in Parliament by Newton, he declined the honor, citing his lack of political qualifications.

■■■■■■■■■■■■■■■■■■■■■■■■■■■■■■■■■■■

Finally, Einstein's reality is a reality in which absolute motion or absolute rest are impossible to determine or define. Those concepts have been obsolete since 1905. All motion, according to new or modern physics, is relative. The position, speed, and direction of anything can only be stated in terms of its relationship to something else.

This something else is generally an observer. It's the observation of position and motion—as well as of reality itself—that determines its nature and value. Each of us, according to Einstein, has our own frame of reference within which, and only within which, motion, position, and reality in general are truly meaningful. And each frame of reference is theoretically just as valid as any other. In this regard modern physics is truly a human, and even personal, doctrine. What exists is what seems to exist.

Starting from the premises of a constant speed of light, a dynamic force field of space, and a completely relativistic definition of motion, modern physics has redesigned the reality of its classical predecessor. The revisions made to classical physics are neither noticeable nor particularly important regarding ordinary events, those occurring within the realm of perception encompassed by the human nervous system. But the farther one explores beyond the bounds of everyday experience the more important the differences between Newtonian and relativistic physics become. At the atomic and cosmic frontiers of physical reality these differences are so profound that they render classical physics useless.

In those realms where Newtonian physics fails, relativistic physics produces some pretty novel, incredible, and just plain bizarre results. While the cause of a few of these can be illustrated with familiar examples—and attempts to do so follow—most of them must be taken on faith alone by anyone who's not a physicist. Because relativity holds the speed of light to be inviolate under any set of circumstances, all other values are required to be flexible in order to accommodate it. All formulas must twist and turn around the ever-constant c. Things formerly thought to be fixed—things like space, time, and mass—become variable.

Newton assumed that "absolute space, in its own nature, without regard to anything external, remains always similar and immovable." According to Einstein, though, space is a subjective phenomenon with a value that seems to change as an observer's velocity changes. As an observer's velocity increases in relationship to that which is being observed, space seems to shrink. A yardstick, moving in relationship to an observer, appears shorter than a stationary yardstick. This effect only becomes noticeable at speeds in the neighborhood of c, and then increases to the point where at the speed of light all objects, regardless of their at-rest lengths, appear to contract into one-dimensional lines.

Newton also assumed that "absolute time, and mathematical time, in and of itself and by its own nature, flows uniformly, without regard to anything external." For Einstein time, like space, is a subjective phenomenon with a value that seems to change as does an observer's velocity. As an observer's velocity increases in relationship to that with which she's measuring time—a clock, a radioactive decay process, the aging of a loved one—time is said to dilate, or slow down. A rapidly moving observer's wristwatch, moving along with her, will seem to mark off seconds, minutes, and hours that are shorter than those being marked off by a rapidly receding watch.

For Newton matter possessed a definite and unalterable mass upon which his laws of motion and gravitation as well as his whole concept of reality depended. For Einstein mass, like space and time, appears to change depending upon the relative velocities of the mass under observation and the observer. As their velocities regarding one another increase the mass of the matter involved seems to increase as well. Again, this phenomenon is only noticeable at speeds in the vicinity of c.

The speed of light, c, plays a very special role in the theory of relativity. Not only is it the single physical constant in the universe, but it's also the fastest speed possible. Nothing, according to the theory of relativity, can travel faster than c. At c, space, which contracts as velocity increases, disappears altogether. At c, time, which slows with velocity, stops completely. At c, mass, which increases along with velocity, becomes infinite.

V.8 | MICKEY AND MINNIE

As its name suggests, the theory of relativity holds that everything, except the speed of light, is relative. The value of everything except the speed of light depends upon an observer's point of view. Two, ten, or a hundred different observers, each in motion relative to the others, will never agree on anything except the value of c. Mass, motion, space, and time, along with all they determine and imply, will be experienced differently by each.

This is because each and every one of us experiences the world about us from our own unique location in space and time. This location determines what sort of messages we receive from physical reality. These messages can never reach us faster than c, the speed of light. No two observers can ever coordinate their points of view any faster than the speed of light allows.

Let's see what sorts of strange things can happen to our perceptions of time and space when the speed of light comes into play. In order to make our illustration more dramatic we're going to exaggerate things a bit. We're going to say that light, instead of traveling at c, ambles along at a mere 1 kilometer per second. By thus slowing light we can magnify the distortions involved in measuring events as they occur and as they're experienced by various observers.

Our relativity experiment involves two participants: Mickey and Minnie. Minnie has just purchased a new motor scooter and wants to find out how fast it will go. Not trusting the manufacturer's speedometer, she devises an experiment. She knows a stretch of smooth, straight highway that has been accurately measured off with markers every kilometer. One quiet night she and Mickey take her scooter to this spot. They each have with them an accurate stopwatch, which they've tested against one another and found to agree. Mickey remains at marker zero while Minnie takes off at full throttle down the road. The plan is for Minnie to flash her lights each time she passes a kilometer marker. She'll note her elapsed time and so be able to calibrate her speed. Mickey will do the same from his vantage point as a double check.

After exactly 1 minute has elapsed on her stopwatch Minnie passes the first marker. She flashes her lights to signal this event to Mickey. The flash, traveling at the speed of 1 kilometer per second for our illustration, takes 1 second to cover the distance back to Mickey. He sees it 1 second after she sends it—1 minute and 1 second after her departure according to his stopwatch. This process continues as Minnie barrels down the asphalt. Every minute, according to her watch, she passes a marker and relays the news to Mickey. Mickey, meanwhile, continues to receive these bulletins at 1-minute and 1-second intervals on his watch. Each flash originates 1 kilometer farther away from him than the previous one and so requires an extra second of travel time to reach him.

When Minnie returns to pick up her partner she's pretty excited. Her new scooter, she figures, can go exactly 1 kilometer every minute, or 60 kilometers per hour. "Just settle down," say Mickey. "Your flashes were 1 minute and 1 second apart. If you flashed at every marker then your scooter needs slightly more than a minute to cover a kilometer. I figure its speed at around 59 kilometers per hour."

"That just can't be, Mick," she counters. "I know I flashed at exactly 1-minute intervals. But your stopwatch has measured the time between them at 1 minute and 1 second. It must be running fast. It's gaining a full second every minute." "Maybe yours is slow," counters Mickey. They check the stopwatches against one another and—lo and behold—find that they still run in perfect agreement.

Distance, like speed and time, is also a bone of contention between them. To Minnie, who's sure her scooter goes 60 kph, a kilometer is the distance she can cover going flat out in one minute. Mickey is confident that his girlfriend's scooter is capable of only 59 kph. His definition of a kilometer is the distance she can cover in 1 minute and 1 second. The distance she goes in a minute is short of a kilometer in his eyes.

What's happened here? One scooter, one highway, and one event have produced two velocities, two different distances, and a difference of opinion between two identical stopwatches. Relativity has struck. There's been but one event, but two observers. Minnie's scooter travels faster from her vantage point than from Mickey's. Minnie sees Mickey's watch as having run fast while Mickey thinks Minnie's ran slow. Mickey's kilometer is longer than Minnie's.

Who's right and who's wrong? According to Einstein, both are equally correct. The speed of the scooter, the performances of the identical stopwatches, and the length of the kilometer are all relative and appear slightly altered from varying perspectives.

To those stuck in the mechanistic mind-set this is all just a clever trick and in no way relevant to the objective reality that exists independently of any observer. The kilometer, the scooter, and the watches, an adherent of classical physics will say, remain the same regardless of who's viewing them or from where.

Not so, says relativity. There is, it maintains, no such thing as a totally objective reality. A kilometer, a scooter, or a watch, in order to be real and meaningful, must be experienced, measured, and discussed by someone someplace. Furthermore, there is no supreme, authoritative someone who has the best vantage point or the final say-so. Each vantage point is equally valid, and each observer has his or her own special perspective.

V.9 | CONFIRMING RELATIVITY

The theory of relativity is full of all sorts of other seemingly crazy ideas besides the distortions of time and distance experienced by Mickey and Minnie. Although these are impossible to illustrate in any common terms, they are all inevitable consequences of a universe in which c is a constant and the fastest speed possible.

No doubt the best known of these is the interconvertability of matter and energy. Matter, according to relativity, is solely pent-up energy; energy solely liberated matter. The two basic components of classical physics are but separate aspects of a single commodity according to modern physics.

Gravity, the driving force behind Newton's mechanistic universe, has also been redefined. It's no longer a mysterious force reaching across empty space, but a function of the nature of space itself. Space, according to Einstein, isn't uniform, but irregular. What we interpret as motion driven by gravity is actually motion driven by the variable consistency of space. Space bends when near a concentration of mass and so relocates everything contained within it. Light rays, too, are bent along with the space through which they travel.

These and other radical ideas of relativity haven't been accepted without resistance. As was Newton before him, Einstein was greeted with ample amounts of skepticism, disbelief, and even ridicule. From the moment he announced his view of the world there were those who, tied to the past and reluctant to change, tried to discredit it. There were others, though, who, able to share his vision, sought to confirm it. To date it has been the pro-Einstein camp that has enjoyed the greater success. Most of this modern genius's predictions have now been verified by either experimentation or observation. None have been contradicted.

The first evidence confirming relativity was found immediately. Newtonian physics had explained the orbits of the moon, the earth, Jupiter, Mars, Uranus, Saturn, and most of the rest of the known universe remarkably well. It had even enabled astronomers to predict the existence of Neptune and then find it based on the effects it had on other planets. But the orbit of Mercury, the planet nearest the sun, had never conformed to the path assigned to it by the law of universal gravitation. The troublesome flaw was completely baffling. Up until 1905, that is. Then, using Einstein's concept of irregular space to explain gravity, the observed orbit agreed with the theoretical one.

A second confirmation came a few years later. Einstein had declared that light, like space, was distorted by the presence of mass. In order for this predicted effect to be noticeable, however, the mass involved has to be enormous. The only familiar object massive enough to bend light significantly, in fact, is the sun. But the sun's brightness ordinarily obliterates any nearby light rays and makes them impossible to examine. In 1919, though, a solar eclipse momentarily darkened the sun and allowed scientists to test the bending of light by mass. The Royal Society of London sent expeditions to Brazil and New Guinea to chart the positions of stars whose light skirted the edges of the sun during the eclipse. These observations were then compared to sightings of these same stars when the sun wasn't a factor. Just as Einstein had predicted, the mass of the sun had bent light and caused a displacement of the stars in much the same way that the surface of a swimming pool causes the displacement of an underwater quarter.

The famous interconvertability of matter and energy was first detected in 1932. In that year it was discovered that cosmic rays, a high-energy form of electromagnetic radiation, sometimes turn into electrons and protons—subatomic bits of matter—as they collide with the atmosphere. The effect was subsequently reproduced in the laboratory. Eventually the reverse process, the conversion of bits of matter into energy, was mastered. The interconvertability of matter and energy, first postulated on Einstein's blackboard, has since been used to, among other things, destroy cities, power submarines, and generate electricity and has solved the mystery of what makes the sun shine. (See Units II.31–32.)

In 1938 it was determined that hydrogen atoms in a state of rapid motion vibrate more slowly than hydrogen atoms at rest. Time, according to these little atomic clocks, slows as the relative speeds of measurer and measuree increase. Einstein had said it would. Newton would have thought the idea pretty peculiar.

Then, in the 1950s, using atom smashers, physicists measured the masses of rapidly moving particles and found them to exceed the at-rest masses of the very same particles. Mass increased along with speed. Relativity again predicted and accounted for this phenomenon whereas classical physics didn't.

And, so it would seem, gravity really is a property of irregular space. Mass really does distort space. Matter and energy really are two sides of the same physical coin. Time really does slow and mass really does increase as velocities approach c. We really do live in a physical world better described by relativity theory than by Newtonian physics. The theory of relativity explains things that classical physics doesn't. It extends the seventeenth-century model of reality into areas then unknown and unaccounted for. Within more mundane areas it's always at least as accurate as its forerunner and usually more so.

For most of us, the relativistic model of reality, despite its fidelity, isn't an easy one to understand. In a few complex equations laden with difficult symbols it contains a whole philosophy. It's a philosophy that staggers the mind and tests the imagination. It's a philosophy that transcends everyday thought and stretches into the realm of pure mathematics. Yet, to those capable of grasping it, it's a beautiful philosophy and, in many regards, a completely logical and surprisingly simple one. Like the moving earth of Copernicus, Einstein's relativity isn't so much complicated as it is novel.

But, like a moving earth, relativity can be conveniently ignored. Few if any of us give much thought to the fact that the earth beneath our feet is spinning and swirling through space. We lead our lives as if terra firma were indeed *firma* (solid), which for all practical purposes it is. Similarly, little if anything we experience in our daily lives is noticeably affected by the implications of relativity. We don't deal with things moving at speeds anywhere near c. We deal with matter in motion at speeds where the theories of relativity and classical physics are virtually identical.

Still, the theory of relativity is important. Its usefulness to those dealing in subatomic physics and the exploration of space—two frontiers that are bound to have an ever-greater impact on everyone—is immeasurable. Its interest to those trying to solve the puzzle of reality is immense.

V.10 | CURVED SPACE

There have been other twentieth-century developments that have resulted in either major revisions or significant additions to the mechanistic model of reality. These, like relativity, haven't had much impact on our daily routines or become part of the common consciousness. Like relativity, though, they've enabled physicists exploring the depths of matter and the fringes of space to make both theoretical and practical progress. They've enabled engineers to design a whole new generation of technology that will transform today's science fiction into tomorrow's true-life adventure. And, perhaps most importantly, they've enabled philosophers to better evaluate the nature of the universe and more firmly fix our place within it.

The first of these additional embellishments to classical physics is closely related to relativity. Up until Einstein unveiled his model of reality space was imagined as an endless sea—a great expanse of either ether or emptiness depending upon whom was asked. Within this sea, it was assumed, satellites orbited, planets revolved, and galaxies swirled. Within this infinite ocean resided the universe as it was known.

The universe itself, it was felt, unlike the sea in which it floated, was of a specific size and shape. It was finite and possessed edges beyond which existed no more galaxies or stars but only unoccupied space reaching out forever. Through various observational techniques it was discovered that this finite universe was expanding. The islands of matter were drifting apart in three dimensions to occupy ever-vaster volumes of space. Perhaps the universe would expand forever or perhaps it would one day begin to contract. In either event, it was assumed that space itself wouldn't be affected—it would remain between and beyond the chunks of matter it contained. It could easily accommodate any future growth and it would remain stoically behind after any contraction.

Infinity, of course, is a difficult if not impossible thing to conceptualize. Still, the only logical way to consider space within the rules of classical physics was as an unbounded and geometrically uniform expanse. If it weren't infinite it would have to have an edge. And if it had an edge then there would have to be something on the other side of that edge. But what? It was easier just to accept an infinity of space surrounding a finite universe.

Einstein changed this attitude in 1905 when he changed the very nature of space. Space, according to relativity theory, isn't just that part of the universe left empty by a shortage of matter, but is a physical entity in its own right. Specifically, it's a four-dimensional space-time continuum. This continuum

■■■■■■■■■■■■■■■■■■■■■■■■■■■■■■■■■■

Finite, but Growing

Finite but unbounded space-time is perfectly compatible with an expanding universe. Einstein was well aware of this fact when he formulated his theory of relativity.

For whatever reasons, however, Einstein was a proponent of a steady-state universe—one that neither expands nor contracts—and set up his equations to yield such a solution. In order to accomplish this goal he had to insert what he called a "cosmological constant." This glorified fudge-factor acted as a sort of universal starch and prevented the otherwise inevitable collapse of the universe under the force of gravity.

Eventually, of course, this bit of chicanery was discovered and the true expanding-universe solution to relativity was worked out. Faced with the facts, a chagrined Einstein fully admitted his duplicity. He later called the cosmological constant the biggest mistake of his life. It's now commonly understood that it's the outward expansion of the universe that counteracts gravity and prevents its collapse.

■■■■■■■■■■■■■■■■■■■■■■■■■■■■■■■■■■

isn't mathematically uniform but is subject to inconsistencies and irregularities. It's distorted by matter.

Based on recent evidence, it's now felt that the space-time continuum, even without any matter twisting it out of shape, is naturally somewhat curved. Space, and the time welded to it, bends, and with it bends everything passing through or transmitted by it: matter, magnetism, gravity, heat, and light. Space-time, instead of extending straight out in all directions equally and forever, gradually sweeps around and eventually envelopes itself. The four-dimensional continuum, while certainly immense, isn't infinite but self-contained. It has a definite size. At the same time though, because it's continuously curved, it lacks an edge. It is, in the jargon of physics, finite but unbounded.

It may seem odd at first that something could be both finite and unbounded. How can space-time have a definite size but no edges? The situation is difficult to imagine only because it occurs in four dimensions—a realm in which we're unaccustomed to thinking and unable to visualize. To demonstrate just how feasible it is for something to be finite yet unbounded let's roll the problem back a few dimensions. Imagine a circle drawn on a piece of paper. Easy, right? Well, the line defining that circle is a bit like curved space—it's finite but unbounded. It has a very specific size or length, namely the circumference of the circle, but no beginning or end. It just happens to exist in two rather than four dimensions. An ideal three-dimensional example of the same situation is the globe we live on. Its surface has a very fixed area yet has no edges. It, too, is finite but unbounded. It possesses a fixed surface area yet lacks edges.

According to the curved-space model, apparently linear space eventually encloses itself to form a finite universe just as the apparently flat surface of the earth eventually wraps around to form a sphere. Just as Columbus sailed over the horizon but remained confined to the earth's surface, a rocket fired out into space will disappear into the distance but remain always within the limits of the universe.

There are, undoubtedly, all sorts of things that exist within the curved space-time continuum yet lie outside our reach. Because he couldn't escape gravity, Columbus was limited to the two-dimensional surface of the globe and prohibited from visiting the moon. Because we can't escape time we're trapped in three dimensions and forbidden from exploring much of our four-dimensional universe.

V.11 QUANTUM MECHANICS

Another development that has resulted in a new outlook on reality has taken place not on the scale of the infinite, but rather of the infinitesimal. It too has had important if subtle practical implications and rather dramatic theoretical ones. And, like relativity, it's a consequence of the ever-closer scrutiny of light.

Light, according to Newton, consisted of countless tiny particles zipping along in straight lines at incredible speeds. It was envisioned as a stream of nearly weightless pieces of matter resembling a spray of water droplets from a high-pressure hose. Certain experiments tended to confirm this strictly mechanistic concept of light, and some, if not all, of its behavior could be faithfully explained in these terms.

Within Newton's lifetime, however, this model of light was challenged by the Dutchman Christian Huygens, who claimed that light consisted not of tiny particles, but of waves of pure energy. This wave theory, too, was supported by certain experiments and some, if not all, of light's behavior could also be explained by Huygens's model.

The debate between proponents of the particle theory and proponents of the wave theory raged for over a hundred years without producing a clear winner. First a major new piece of evidence would support the wave theory and it would gain favor. Then another discovery would support the particle theory and it would take precedence. This seesaw battle was one of the liveliest ever to be waged in the arena of physics. It was also one of the most frustrating, for no matter how much of light's behavior one theory explained, there were always certain aspects of it that could only be explained by the opposing theory. The true and total nature of light defied all attempts to describe it in terms of either model.

What was needed, of course, was a whole new approach, one that, like relativity, resided outside the parameters of classical physics. Based on the pioneering efforts of the German Max Planck (1858–1947) in 1900, the contributions of Einstein in 1905, of the Dane Niels Bohr in 1913, and another 20 years or so of hard work by many others, a radical new model of light was eventually constructed. It's known now as the quantum mechanistic model.

Although extremely complicated and, once more, expressible and understandable only in the language of mathematics, the crux of the quantum mechanistic model is that light exists neither as particles nor as waves, but as something resembling both but different from either. That something is the quantum. A quantum is a particle-wave hybrid acting sometimes like one, sometimes like the other, but never as both simultaneously. It's been dubbed "a packet of energy," "a wave pattern of particles," and, most cleverly, "a wavicle." The terminology really doesn't matter. What matters is that light has a dual nature, a nature that's uniquely its own and one that can't be discussed or interpreted in ordinary terms. Like relativity and curved space, it's impossible to simplify it to anything other than what it is.

Quantum mechanics extends beyond the scope of visible light to include all forms of radiant energy: infrared heat, ultraviolet light, x-rays, microwaves, radar, and the rest. Electromagnetic energy in general, according to this model, doesn't exist as a smooth unbroken flow, but as a collection of individual units. Each unit is a quantum, also called a photon within the range of visible light. Radiant energy then, like matter, is discontinuous. The quantum is to energy what the atom is to matter—the smallest unit into which it can be meaningfully divided. Energy can exist only in even multiples of quanta, never in fractional amounts.

At the level of human perception, of course, the discontinuity of energy is of no consequence. We can just as safely go about our business oblivious of the existence of quanta as we can oblivious to the existence of atoms. We deal only with events involving astronomical numbers of quanta just as we deal only with objects consisting of astronomical numbers of atoms. When acting en masse, quanta, like atoms, lose their individual identities. Their collective behavior is what generally concerns us and what is described by classical physics.

The behavior of individual quanta, however, doesn't follow the same rules as does their collective behavior. Events involving only a few atoms or quanta don't obey the laws of Newtonian physics. At the submicroscopic level a whole different physics applies—the physics of quantum mechanics.

V.12 THE UNCERTAINTY PRINCIPLE

The most significant difference between quantum and classical mechanics, between new and old physics, is their degree of predictability. The Newtonian model of reality is grounded in certainty. A body with X mass, subjected to Y force, is accelerated in the amount of Z and will be every time. An apple will always fall identically according to the law of universal gravitation. Given the magnitude and direction of any action the size and direction of the ensuing reaction are utterly and precisely predictable. The law of cause and effect operates without exception in the macroscopic world.

In the microscopic world it doesn't. Because the energy that powers them is discontinuous, events at the atomic and subatomic levels take place not as fluid actions but as a series of disjointed jerks and jumps. A quantum-sized event either takes place completely or not at all. There's no middle ground, no gradual progression. Reality in this sense is something like a strip of movie film. It proceeds from one static frame to the next in a series of small increments. Each frame differs slightly from the one that preceded it and from the one to follow. The apparent smoothness is only an illusion. (See Figure V.1.)

During each quantum of change, as between each frame of a movie, there's a moment of suspense. One state of affairs has become history and another is about to be realized. During these momentary gaps in the flow of reality it's impossible to say just what the next state of affairs will be, for quantum mechanics, unlike its classical counterpart, is incapable of certainty. The future is describable only in terms of probability. Given an existing set of circumstances a future event has nothing more than a statistical likelihood of happening.

On the macroscopic-collective scale we can state with total confidence that one-half the atoms in any given sample of polonium-210 will decay into atoms of lead in 139 days. But, at the atomic-quantum scale we have no way of knowing for sure which atoms will decay, which won't, or when any particular atom of polonium will transmutate into lead. Macroscopic events, involving millions of atoms and quanta, derive their certainty only from the averaging of their components. Microscopic events, confined to a few specific atoms and quanta, are impossible to predict accurately. Reality as we experience it derives its predictability from what's known as the law of large numbers. It is, in fact, not certain at all, but only highly probable. It's a generalization based on the average results of many, many, many individual events. Each individual event is unique, unpredictable, and unconnected to the others. Together they produce the illusions of continuity, certainty, and causality.

When the statistical nature of microscopic reality was first discovered it caused no great commotion. Everyone assumed, understandably enough, that the lack of certainty involved was due only to a lack of sufficient data. A coin toss is a random event unless one knows all there is to know about the mass of the coin, the force applied to it, its rotational momentum, the strength and direction of the wind, the distance it will fall, etc. Then it becomes a calculable certainty. It was widely assumed that, given enough measurements, quantum mechanics would eventually be reduced to a few concrete laws. Even Einstein was unable to accept the fact that reality proceeded with a certain amount of

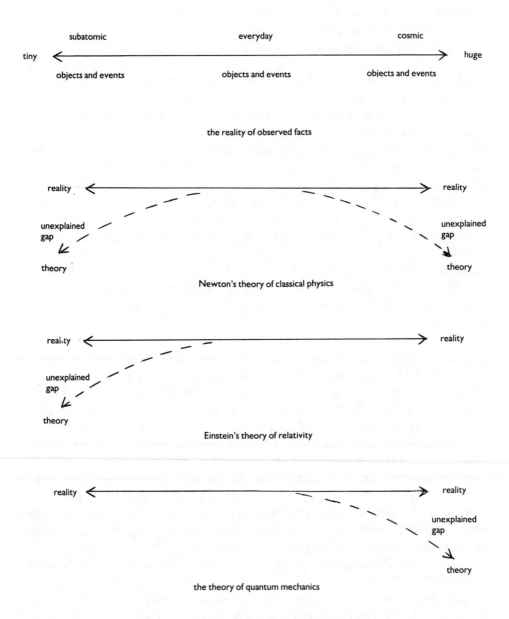

Figure V.1   Comparison of theories to reality

randomness. "I can believe that God governs the world according to any set of laws," he said, "but I cannot believe that God plays dice."

Then, in 1927, a German physicist, Werner Heinsenberg (1901–1976), first stated what has since been known as either the theory of indeterminacy or the uncertainty principle. What Heisenberg specifically discovered was the fact that it's impossible to accurately measure both the position and the momentum of a subatomic particle simultaneously. The very act of measuring position, he found, disturbs the particle's momentum, and vice versa. The implications of this superficially modest and esoteric finding are immense: perfect knowledge of events on an atomic scale is theoretically unattainable.

Determinism, according to the uncertainty principle, is a sham. There's no exactness, only a certain degree of approximation. There's no absolute truth, only a best guess. This doesn't necessarily mean that God plays dice; it only means that whatever it is he, she, or it does play is ultimately unknowable to us as human beings.

V.13 | A MATTER OF OPINION

Documented facts interpreted by the scientific method indicate that the universe began in a cataclysmic explosion some 17 billion years ago. Affectionately known as the Big Bang, this primal fireball consisted of massive amounts of energy and little else. Some of the energy subsequently condensed into matter according to Einstein's formula. The shock wave of the explosion spread outward carrying energy and the newly created matter along with it. Space was created and rapidly expanded. Energy and matter began to interact. There was motion. There were events. There was change. There was time.

As the universe continued to inflate, the bits and pieces of matter collided, stuck together, and became simple atoms of hydrogen and helium. These atoms, attracted to one another by gravity, coalesced to form clouds of gas and dust. The clouds, again powered by gravity, condensed further to form stars. Some massive stars, acting as giant nuclear furnaces, fused the light elements into heavier, more complex ones like iron and nickel. The stars then exploded and spewed these new forms of matter into space. As explosions and condensations continued to alternate, the matter assumed the form of galaxies, solar systems, planets, and satellites. This evolutionary process continues to this day and can be seen in all its multiple stages with the aid of powerful telescopes.

Our own solar system was formed about 4.5 billion years ago. Our planet, favorably poised between the searing heat of the sun and the frigidity of space, stabilized at a temperature that allowed for the existence of liquid water. Ultraviolet rays from the sun triggered chemical reactions among the simple atoms and molecules. From water, carbon dioxide, and oxygen arose hydrogen cyanide and formaldehyde, then amino acids, nucleic acids, sugars, and proteins. These complex molecules, driven by a constant input of solar energy, then organized themselves into the simplest forms of life. Urged on by a will to survive, life became more and more complex. Finally, after millions of years of effort and repeated failures, it evolved into *Homo sapiens*. With this accomplishment the Big Bang produced a consciousness capable of contemplating itself, its origin, and its situation.

With an awareness of reality came a need to know what it was. The initial response was one of utter amazement. Reality was a miracle, our existence a mystery, and our fate out of our hands. We were helpless, accidental inhabitants of a universe that was completely beyond approach. But later, gaining information and, along with it, confidence, attempts were made to sort out the various aspects of the physical world. Schemes were proposed that organized what had been thought of as random events into categories and patterns. With the perfection of the scientific method these schemes became a matter of fact rather than opinion. Work began on constructing a model of reality that was agreed upon by consensus and objective evidence.

This model has been under construction for over 300 years now and has resulted in some spectacular achievements. We've tailored matter according to our desires. We've reproduced the nuclear blast of the stars here on earth. We've visited the moon. We've created whole new forms of life.

The model that has allowed this headlong rush of progress is basically the mechanical one of Newton modified and enlarged by relativity and quantum mechanics. It's called, more or less officially, the relativistic quantum field theory. Like its predecessors, the relativistic quantum field theory is only an approximate description of reality. Not perfect, but the best we've been able to come up with so far. It's sufficiently accurate to answer a vast array of questions. We know how planes fly, where wood goes when it burns, and what makes the sky blue. But other questions remain unanswered. What is matter? What is energy? What is space? What is time? What is reality?

These questions, apparently, aren't answerable individually. Matter and energy are different formats of the same maddeningly elusive stuff. Space and time are different aspects of the same four-dimensional force field. Even existence and nonexistence are statistical manifestations of a mathematical abstraction. There are no fundamental parts from which reality is built. It's not an assembly of discrete components but rather a single, integrated oneness. The fragmentation inflicted upon it for the sake of analysis is an artificial device—a product of human convenience—and no fraction of reality thus created can ever be satisfactorily understood in isolation.

But it's now thought that not only are these questions unanswerable individually, they're unanswerable in any absolute sense. Any answers that may exist are totally relative. Just as everyone in a football stadium sees a touchdown from a slightly different perspective, we all experience reality from inside our own heads. We all must find our own answers. No two of these answers need be identical. None is supreme. Reality is, as much as anything else, a matter of opinion. "Physical concepts," as Einstein put it, "are free creations of the

■■■■■■■■■■■■■■■■■■■■■■■■■■■■■■■■■■■■

A Theory of Everything

There have been two major revisions of Newton's classical physics in the twentieth century. The first of these—the theory of relativity—deals with corrections needed on the scale of the superlarge. The second—quantum mechanics—deals with corrections needed on the scale of the supersmall. Science is a dynamic discipline, and these modifications are seen as an inevitable part of the constant fine-tuning of theory to reality.

What is troublesome, though, is the fundamental incompatibility of the laws of relativity with those of quantum mechanics. To scientists, whose foremost goal is to simplify things, this duality is intolerable. The unification of these two divergent areas is now seen as the next and, some say, final goal of theoretical physics.

Einstein struggled to bridge this gap for the last 25 years of his life, but never found a solution that satisfied him. Current experts are confident that a solution exists, however, and some have even predicted it will be found within the very near future. The Nobel Prize awaits.

■■■■■■■■■■■■■■■■■■■■■■■■■■■■■■■■■■■■

human mind, and are not, however it may seem, uniquely determined by the external world." Reality exists in us as much as we exist in it.

Finally, the answers, or answer, to the tough questions, even for any particular observer, are ultimately unknowable. Matter only "tends to occur" at certain locations. Events only "tend to occur" at certain times. Reality is a fragile, untouchable thing. The very act of observing disturbs that which is observed. There's no unbroken string of cause and effect, only a long chain of probabilities. We and our world exist, most likely, as a product of pure chance, not grand design.

Science has come full circle. The reality that it first assumed was distant and unknowable it now regards as intimate and highly subjective. The laws that it laid down to describe the physical world are now said to describe nothing more than our consciousness. Sound is what we hear. Light is what we see. Reality is what we experience. The external world is a product of the internal world.

The ultimate solution to the puzzle of reality, then, won't be found in the atom, the laboratory, or the distant galaxies, but rather in the heart, mind, and soul. The quest handed down from the priests to the philosophers, and then from the philosophers to the scientists, is now being passed back along the same route.

SCIENCE AND TECHNOLOGY: A BRIEF HISTORY

Copernicus — moving earth	1540	Vettelli — pistol
	1550	unknown — screwdriver
	1560	Gesner — pencil
	1570	
Galileo — pendulum motion	1580	
Pope Gregory XIII — calendar		
Galileo — falling bodies	1590	Janssen — microscope
Gilbert — laws of magnetism	1600	
Galileo — moons of Jupiter		Lippershey — telescope
Kepler — *New Astronomy*	1610	Harvey — circulation of blood
	1620	Oughtred — slide rule
	1630	Descartes — analytical geometry
Mersenne — speed of sound	1640	
Torricelli — barometer		Pascal — hydraulic press
	1650	
Boyle — *The Sceptical Chymist*	1660	Huygens — pendulum clock
Newton — universal gravitation		Newton — integral calculus
	1670	
Roemer — speed of light	1680	Papin — pressure cooker
Newton — *Principia*	1690	Papin — steam engine
Stahl — phlogiston	1700	Leibniz — binary arithmetic
	1710	Shore — tuning fork
Fahrenheit — thermometer	1720	
Bradley — aberration of light	1730	Kay — flying shuttle
	1740	

	1750	
Black — carbon dioxide		Franklin — lightning rod
	1760	
Cavendish — hydrogen		Hargreaves — spinning jenny
	1770	Watt — steam-engine condensor
Priestley — oxygen		
	1780	Watt — copying machine
	1790	Saint — sewing machine
Lavoisier — elements		
Thompson — heat from friction		
Volta — voltaic pile	1800	Whitney — cotton gin
Dalton — atomic theory		
	1810	Donkin — canned food
Oersted — electromagnetism	1820	
		Stephenson — steam locomotive
Faraday — induction	1830	
		Morse — telegraph
Joule — conservation of energy	1840	Daguerre — photography
Doppler — Doppler effect		
	1850	Harrison — refrigerator
Mendel — genetics		
	1860	Darwin — *Origin of Species*
		Nobel — dynamite
Mendeleev — periodic chart	1870	Hyatt — plastic
		Bell — telephone
	1880	Edison — incandescent lightbulb
railroads — time zones		Daimler — gas-powered auto
	1890	
Roentgen — x-rays		Marconi — wireless communication
Curies — radioactivity	1900	
Einstein — relativity		Wright brothers — first flight
	1910	Ford — moving assembly line
Bohr — atomic structure		
	1920	Johannes — AM radio
		Baird — television
Heisenberg — uncertainty		
Born — quantum mechanics	1930	Fleming — penicillin
Jansky — radio astronomy		Carothers — nylon
Fermi — chain reaction	1940	Goldmark — color television
Oppenheimer — atomic bomb		Yeager — supersonic flight

Libby — carbon-14 dating
Teller — hydrogen bomb
Townes — laser
Schmidt — quasars
Boyd — black holes
U.S. Army — neutron bomb

Sony — compact disk

1950   Shockley — transistor
        Salk — polio vaccine
1960   Gagarin — earth orbit

1970   Armstrong — moon walk
        NASA — space shuttle
1980   Steptoe — test-tube baby
        Jarvik — artificial heart
1990

SELECTED BIBLIOGRAPHY

If you need or want more information on any of the topics discussed in *A Handbook to the Universe,* the textbooks and encyclopedias listed here can probably provide it. They are all widely available in school and city libraries. The books suggested under Supplementary Reading offer a taste of the fine science writing that is readily available in libraries and bookstores. Don't hesitate to ask your teachers or librarians for more suggestions, for magazines as well as books. This Bibliography is just a beginning.

Physical Science Textbooks
for grades 6–9:

Bacher, Hurd, McLaughlin, Silver. *Prentice Hall Physical Science.* Englewood Cliffs, NJ: Prentice Hall/Simon & Schuster Educational Group, 1993.

Carle, Sarquis, Nolan. *Physical Science: The Challenge of Discovery.* Lexington, MA: D. C. Heath and Company, 1991.

Cooney, J. Pasachoff, N. Pasachoff. *Scott Foresman Physical Science.* Glenview, IL: Scott Foresman, 1990.

DiSpezio, Linner-Luebe, Lisowski, Skoog, Sparks. *Science Insights: Exploring Matter and Energy.* Menlo Park, CA: Addison-Wesley Publishing Company, 1994.

Eby and Horton. *Macmillan Physical Science.* Columbus, OH: Macmillan/McGraw-Hill School Publishing Co., 1988.

Smith, Thompson, McGervey, Ballinger. *Merrill Physical Science.* Columbus, OH: Glencoe Macmillan/McGraw-Hill, 1993.

for high school:

Tropp and Friedl. *Modern Physical Science.* Orlando, FL: Holt, Rinehart and Winston, Inc., 1991.

Chemistry Textbooks
for high school:

Dorin, Demmin, Gabel. *Prentice Hall Chemistry: The Study of Matter.* Englewood Cliffs, NJ: Prentice Hall/Simon & Schuster Education Group, 1992.

Herron, J. Sarquis, M. Sarquis, Schrader, Kukla. *Heath Chemistry.* Lexington, MA: D. C. Heath & Company, 1993.

Smoot, Smith, Price. *Merrill Chemistry.* Columbus, OH: Glencoe Macmillan McGraw-Hill, 1993.

Tzimopoulos, et al. *Modern Chemistry.* Orlando, FL: Holt, Rinehart and Winston, Inc., 1993.

college chemistry for liberal arts students:

Joesten, Johnson, Netterville, Wood. *World of Chemistry.* Orlando, FL: Saunders College Publishing, a division of Holt, Rinehart and Winston, Inc., 1991. This text is correlated with *The World of Chemistry,* Annenberg/CBP Project, 26 thirty-minute video programs.

Physics Textbooks for high school:

Hewitt, Yan, Robinson. *Conceptual Physics.* Menlo Park, CA: Addison-Wesley Publishing Company, 1992.

Martindale, Heath, Konrad, Macnaughton, Carle. *Heath Physics.* Lexington, MA: D. C. Heath and Company, 1992.

Taffel. *Physics: Its Methods and Meanings.* Englewood Cliffs, NJ: Prentice Hall/Simon & Schuster Educational Group, 1992.

Trinklein. *Modern Physics.* Orlando, FL: Holt, Rinehart and Winston, Inc., 1992.

Zitzewitz and Murphy. *Physics: Principles and Problems.* Columbus, OH: Glencoe Macmillan/McGraw-Hill, 1992.

Science Encyclopedias:

McGraw-Hill Encyclopedia of Science and Technology. New York: McGraw Hill Book Company, 6th Edition, 1987. Nineteen volumes of about 650 pages each, plus an index volume.

McGraw-Hill Yearbook of Science and Technology. New York: McGraw-Hill Book Company, 1988 to present. One volume of about 500 pages is released each year.

The Raintree Illustrated Science Encyclopedia. Austin, TX: Steck-Vaugn Company, 1992. Seventeen volumes of about 120 pages each; Volume 18 contains science projects, bibliography, and index.

Van Nostrand's Scientific Encyclopedia. New York: Van Nostrand Reinhold Company, Sixth Edition, 1983. Two volumes totalling 3069 pages.

Supplementary Reading:

Asimov, Isaac. *Frontiers: New Discoveries About Man and His Planet, Outer Space and the Universe.* New York: Truman Talley Books/Plume, 1991.

Carr, Joseph J. *The Art of Science: A Practical Guide to Experiments, Observations, and Handling Data.* San Diego: Hightext Publications, Inc., 1993.

Faraday, Michael. *Faraday's Chemical History of a Candle.* Chicago: Chicago Review Press, Inc., 1988.

Feynman, Richard. *The Character of Physical Law.* Cambridge: The MIT Press, 1967.

Hawking, Stephen. *A Brief History of Time: From the Big Bang to Black Holes.* New York: Bantam, 1990.

Kuhn, Thomas S. *The Structure of Scientific Revolutions.* Chicago: The University of Chicago Press, 1970.

Lederman, Leon. *The God Particle: If the Universe Is the Answer, What Is the Question?* New York: Houghton Mifflin, 1993.

Lightman, Alan. *Ancient Light: Our Changing View of the Universe.* Cambridge: Harvard University Press, 1991.

INDEX

A

Aberration of starlight, 177
A-bomb, 149–50
Absolute zero, 94, 96
Acceleration, 81–82
Acetic acid, 61
Acetylene, 54
Acoustics, 112–14
Adhesion, 51–52
Aerosol sprays, 54–55, 57
Aether, 161, 170, 248, 252
Air, 28, 37, 57–59, 113–15. *See also*
   Atmosphere; Wind
Alchemy, 8
Allotropes, 20–22, 41
Alloys, 29, 49–50
Almagest (Ptolemy), 162–64
Alpha radiation, 63–64
Alternating current (AC), 143
Aluminum, 28
Ammonia, 60
Amperage, 141
Ampère, André-Marie, 141
Anaximenes, 4, 57
Anode, 142
Antarctic Circle, 184
Archimedes, 36, 79–80
Arctic Circle, 184
Aristarchus, 165
Aristotle, 5–6, 57, 79–80, 127, 161–62,
   201, 248
Asteroids, 191–92
Astrology, 158, 163, 196
Astronomia Nova (Kepler), 167
Astronomy, 157–62, 198, 241
Atmosphere, 44, 57–58, 110, 133, 177,
   185, 197–98
Atomic bomb, 149–50
Atomic energy, 144–46
Atomic mass units (amu), 65–66
Atomic number, 19, 65
Atomic theory, 4–5, 9–11, 14–16
Atomic weight, 15, 65–66
Atoms, 4–5, 9–11, 14–16, 64–66, 144
Aurora australis, 137
Aurora borealis, 137

Avogadro, Amedeo, 56
Axis (of Earth), 180–83, 216

B

Baekeland, Leo H., 45
Bakelite, 45
Barometric pressure, 57–58. *See also* At-
   mosphere
Bathtub ring, 61
Batteries, 141–42
Bauxite, 28
Becquerel, Antoine-Henri, 62–63, 144
Bernoulli, David, 59
Bernoulli effect, 59
Berzelius, Jöns Jacob, 16
Beta radiation, 63–64
Bible, 207–8, 210
Big Bang, 197–99, 239, 269
Big Crunch, 241
Black, Joseph, 11
Black holes, 241
Board of Longitude, 240
Bohr, Niels, 64–65, 67, 266
Boiling point, 39, 41, 98–101
Bomb, atomic, 149–50
Bomb, hydrogen, 153
Borazon, 39
Boyle, Robert, 9–10, 112
Boyle's law, 10, 56
Brass, 50
Breeder reactors, 149–50
Brightness, 126
Brittleness, 39
Bronze, 50
Bruno, Giordano, 169
Butane, 32

C

Caesar, Julius, 219–20
Calendars: Julian, 218–21, 223–24; lu-
   nar, 217–19; Gregorian, 218, 223–
   25; Roman, 218–19; solar, 219–20
Caloric, 73, 75
Capillary action, 52
Carats, 50
Carbon, 20–22, 42

Carbon dioxide, 11–14
Carbon-14 dating, 67, 145
Catalysts, 32, 49, 91
Cathode, 143
Catholic Church. *See* Church
Celluloid, 45
Celsius, Anders, 94
Celsius temperature scale, 94–96
Centigrade temperature scale, 94
Chadwick, James, 66
Chain reaction, 146–49
Chamberlain, Thomas Chrowder, 212
Charges, electrical, 138–40
Charles's law, 56
Chemical reactions, 22–23, 26, 65, 90–91, 141–43
Chemical symbols, 16
Chemistry, 29, 31, 66
Christmas, 210, 221, 223
Church, 7–8, 168–69, 207–8, 220–21, 223, 248–49
Clavius, Christoph, 223
Clocks, 226–29, 232
Clouds, 62, 104–5, 130
Cockcroft, John, 145
Cohesion, 51–52
Colloids, 60–62
Color, 126–30
Combustion, 6, 11–13, 90–91
Comets, 191–92
Compasses, 136–37, 240
Compounds, 22–24
Condensation, 102, 104–6
Conduction, 106–9
Conductivity, 4, 39
Conductors, 138–39
Conservation of energy, law of, 75–76
Conservation of matter, law of, 13
Contrails, 105
Convection, 106–9
Copernicus, 158, 163–65, 170, 179, 223, 249, 253
Copper, 28, 49
Coriolis effect, 182
Cosmic radiation, 135, 262
Cracking, 32
Crux, 194
Crystals, 44, 47–48
Curie, Irène, 63
Curie, Marie Sklodowska, 62–63, 144
Curie, Pierre, 62–63
Current, electrical, 139–43

D
Dalton, John, 14–16
Darwin, Charles, 209
Davy, Humphry, 75
Day, 214–16, 225, 229–30
Decibels, 118
Democritus, 1, 4–5
Density, 11, 35–36, 40, 59
Depth perception, 120–21
*De Revolutionibus Orbium Coelestium.
See Revolutions of the Heavenly Bodies*
Detergents, 61
Determinism, 250, 269
Dew, 105–6
*Dialogue Concerning the Two Chief
World Systems—Ptolemaic and Copernican* (Galileo), 168
Diamond, 20–22, 39, 41, 123
Dionysius Exiguus, 221
Direct current (DC), 143
Distillation, 27–28, 31
Doppler, Christian Johann, 119
Doppler effect, 119, 197, 212

E
Earth (planet), 158, 178; age of, 208, 211; as center of universe, 157, 161–63, 168–70; orbit of, 164–65, 174–75, 178–79; origins of, 207–8, 211–12; rotation of, 164–65, 179–82, 230; size and shape of, 160, 178
Earth (substance), composition of, 28–29
Earthquakes, 117
Easter, 210, 223
Echoes, 114
Eclipses, 159–60, 190
Ecliptic, 182, 188, 192
Einstein, Albert, 146–47, 256–57, 264, 266–67, 269–70
Elasticity, 39, 41, 113–15
Electricity, 41, 62, 138–43
Electromagnetic induction, 142, 149
Electromagnetic radiation, 106–9, 122–26, 131–35. *See also* Light
Electrons, 41, 62, 137–43, 144
Elements, 4, 13, 19–20, 24; ancient Greek, 4–7, 248; artificial, 69; discovery of, 9–11, 13, 16; occurrence of, 20; symbols of, 16
Empedocles, 4, 57, 127

Emulsifiers, 61
Energy, 73–78, 89. *See also* Electricity; Heat; Light; Mechanics; Nuclear energy; Sound
Entropy, 206, 239
Epicycles, 161–62, 164, 166
Equator, 180, 184, 194
Equinoxes, 184, 196
Eratosthenes, 160
Escapement, 227–28
Escape velocity, 175
Ethane, 32
Ether, 252–56
Evaporation, 27, 102–4
Exosphere, 58
Extraction, 27

F
Fahrenheit, Daniel Gabriel, 94
Fahrenheit temperature scale, 94–96
Falling bodies, 8–9, 80–83
Fallout, radioactive, 149
Faraday, Michael, 142
Feldspar, 28, 30
Fermi, Enrico, 148
Fission, nuclear, 150–52
Fitzgerald-Lorenz contraction effect, 256
Flight, principle of, 59
Flotation, 27, 35
Fluorescence, 120
Foams, 61
Fog. *See* Clouds
Foot-candles, 126
Force fields, 135–36
Foucault, Jean, 180
Franklin, Benjamin, 126, 140
Freezing point. *See* Melting point
Freon, 98
Fresnel drag coefficient, 256
Friction, 78, 88–89
Frisch, Otto, 146
Frost, 106
Fructose, 23–24
Fusion, latent heat of, 96–97
Fusion, nuclear, 150–53

G
Galaxies, 193–94
Galena, 28, 30
Galileo Galilei, investigations of: air, 57; astronomy, 167–69, 253–54; fall-ing bodies, 8–9, 80–83; heat, 92–93; light, 122, 125; pendulum, 228
Gamma radiation, 135, 144–45, 149
Gas, LP, 32
Gas, natural, 32
Gases, 44, 54–57
Gems, 48
General Time Convention, 235
Generators, 142
Gilbert, William, 136
Glass, 45, 92
Glucose, 23–24
Gods, 159, 246, 251
Gold, 49–50
Granite, 28, 30
Graphite, 20–22, 41
Gravitation, 27
Gravity, 51, 57–58, 174–75, 261. *See also* Universal gravitation
Gravity and levity, ancient Greek, 6–8, 79, 81, 248
Greenhouse effect, 108, 110
Greenwich Standard Time, 236
Gregory XIII (Pope), 223–24

H
Hahn, Otto, 146
Hail. *See* Precipitation
Half-lives, 209–10, 241
Halley, Edmund, 172
Hardness, 38–39, 41
Harrison, John, 240
H-bomb, 153
Hearing, sense of, 117–19
Heat, 77–78; definition of, 91; flow of, 106–10; and friction, 88–89; and expansion, 92; nature of, 89, 206; radiant, 131–33. *See also* Temperature
Heat capacity, 37–38, 40, 53
Heisenberg, Werner, 269
Hematite, 28–29
Heraclitus, 4, 71
Hipparchus, 160
Holy Roman and Universal Inquisition, 168–69
Hour, 225
Hubble, Edwin Powell, 213
Hubble's law, 213
Humidity, 101–4, 139
Hutton, James, 208
Huygens, Christian, 228, 265
Hyatt, John Wesley, 45

Hydrocarbons, 32
Hydrogen, fusion of, 151–53
Hydrogen bomb, 153
Hydrosphere, 44

I
Ice. *See* Water
Ice, dry, 42
Impetus, 80–81, 83
Incandescence, 119–20
Indeterminacy, theory of, 269
Index Librorum Expurgatorus, 168
Inertia, 81–84, 174–75
Infrared radiation, 131–33, 144–45, 198
Infrasound, 117
Insulators, 106, 109, 138–39
Intercalation, 217–18, 231
International Date Line, 237–38
International Earth Rotation Service, 231
International Meridian Conference, 235
Ionosphere, 58
Iron, 28–29, 49, 91
Isomers, 24
Isotopes, 66–69, 147, 209, 211

J-K
Jet lag, 237
Jet propulsion, 87
Jet stream, 58
Joliot, Frédéric, 63
Joule, James Prescott, 75
Jupiter, 125, 158, 168, 191–92, 240

Karats, 50
Kelvin (Lord). *See* Thomson, William
Kelvin temperature scale, 94, 96
Kepler, Johannes, 166–67, 170
Kinetic molecular theory. *See* Matter
Koppernigk, Nicholas. *See* Copernicus

L
Laplace, Pierre-Simon de, 211
Lasers, 67, 153
Latitude, 240
Lavoisier, Antoine-Laurent, 11–13, 73
Lead, 28
Leap days, 231
Leap years, 220, 223, 231
Leclerc, Georges-Louis, 208, 211
Lenses, 123

Leo X (Pope), 223
Levity. *See* Gravity and levity, ancient Greek
Light, 119–23; aberration of, 177; bending of, 261; nature of, 252, 265–66; speed of, 122–23, 125, 256–58. *See also* Electromagnetic radiation
Lightning, 123, 139–40
Light year, 193
Liquids, 42–43, 51–53
Lithosphere, 44
Lodestone, 136
Longitude, 236–37, 240
Loudness, 112
Luminosity, 120
Lyceum, 5
Lyell, Charles, 209

M
Mach, Ernst, 116
Mach speed, 116
Magnetism, 135–38, 153
Magnetite, 28–29, 136
Malachite, 29
Malleability, 39, 41
Manhattan Project, 148
Mars, 158
Mass, 33–34, 40; critical, 148; and gravity, 171–73; and momentum, 85–86; relative nature of, 258, 261–62
Mathematical Principles of Natural Philosophy. See Principia
Mathematike Syntaxis. See Almagest
Matter, 3–6, 27, 151, 241, 269; conversion into energy, 145–46, 151; kinetic molecular theory of, 40, 89; law of conservation of, 75–76; states of, 41–44
Mechanics, 78, 83, 85, 173–75
Meitner, Lise, 146
Melting point, 39, 41, 44, 53, 96–97, 101
Mendeleev, Dmitri, 16–19
Mercury (planet), 158, 261
Meridian, 232
Mersenne, Marin, 114
Mesosphere, 58
Metals, 28–29, 49–51
Methane, 14–15, 32
Meton, 218
Metonic cycle, 218

Michelson, Albert A., 255
Microwave radiation, 132–34
Milky Way, 193–94
Minerals, 28–30
Mirages, 123
Mixtures, 26–27
Molecules, 15, 22–23, 26, 40–44
Momentum, 85–87
Month, 188, 214, 216
Moon, 157, 174–75, 186–91
Morley, Edward W., 255
Moseley, Henry Gwyn Jeffreys, 65
Motion, 203, 253–54, 256; laws of,
    83–85, 171, 173; periodic, 226–
    29

N-O
Neptune, 158
Neutrons, 66
New Astronomy (Kepler), 167
A New System of Chemistry (Dalton), 15
Newton, Isaac, 83–85, 127–28, 170–
    73, 250, 265
Nicaea, Council of, 210
Noise. *See* Sound
Northern lights. *See* Aurora borealis
North Pole. *See* Poles, geographic
North Star, 194
Nuclear bomb. *See* Bomb, atomic;
    Bomb, hydrogen
Nuclear energy, 144–46, 148–50
Nuclear fission. *See* Fission, nuclear
Nuclear fusion. *See* Fusion, nuclear
Nuclear physics. *See* Physics
Nuclear waste, 145
Nucleus, atomic, 64
Nylon, 46

Octane, 24
Oersted, Hans Christian, 142
Ohm, Georg Simon, 141
On the Origin of Species by Means of Natu-
    ral Selection (Darwin), 209
Opaqueness, 129–30
Oppenheimer, J. Robert, 149
Optics, 123
Orbits, 174–76, 261
Ores, 28–30
Oscillating universe, 197–98, 241
Oxidation, 90–91
Oxygen, 11–13, 20–22
Ozone, 20–22

P-Q
Papin, Denis, 101
Parallax, 121, 179, 193
Pendulum, 180, 228
Periodic chart, 16–19, 22–23, 65
Petroleum, 31–32
Philosophers' Stone, 8
Philosophiae Naturalis Principia Mathe-
    matica. See Principia
Phlogiston, 11–13
Photoelectric effect, 120
Photon, 266
Photosynthesis, 90
Physics, 66, 250–56, 266–67
Pitch, 112
Planck, Max, 266
Planets, 157–59, 161, 166–67, 173,
    191–93
Plastic, 45–47
Platinum, 49
Plato, 155
Pluto, 158, 192
Plutonium, 69, 149–50
Polaris, 194
Polarity, magnetic, 136–37
Poles, geographic, 136–37, 194
Polymers, 45–47
Precipitation, 59, 104–5
Pressure, of gases, 55; at boiling and
    freezing points, 100–101. *See also* At-
    mospheric pressure
Pressure cooker, 100–101
Priestley, Joseph, 12
Prime meridian, 236, 240
Principia (Newton), 172, 250
Properties of matter, 6–7, 34–35, 165
Protagorus, 243
Protons, 63, 65, 137–38
Ptolemy, Claudius, 162
Pytheas, 190

Quanta, 266
Quantum mechanics, 266–68, 271
Quartz, 28, 30, 229

R
Radar, 132–33
Radiation. *See* Electromagnetic radiation
Radioactive decay (radioactivity), 62–
    64, 67–69, 144–46, 209–10
Rain. *See* Precipitation
Rainbow, 128

Reactors, nuclear, 149–50
Red shift (of starlight), 212
Reflection, 129–30
Refraction, 122–23
Refrigerators, operation of, 98–100
Relativistic quantum field theory, 270
Relativity, theory of, 256–62, 271
Resistance, 39, 41, 141
Resonance (reverberation), 113–14
Revolutions of the Heavenly Bodies (Copernicus), 164, 168, 249
Rocket propulsion, 85–87
Roemer, Olaus, 125
Roentgen, Wilhelm, 135
Roentgen radiation. *See* X-rays
Roman Catholic Church. *See* Church
Roosevelt, Franklin, 147–48
Rumford (Count). *See* Thompson, Benjamin
Rust. *See* Oxidation
Rutherford, Ernest, 63, 144

S
Saros, 190
Satellites, 173
Saturation, 61
Saturn, 158
The Sceptical Chymist (Boyle), 9
Scientific method, 8–10, 247
Seasons, 182–85
Seawater, 26–28
Sedimentation, 60
Sight. *See* Vision
Silica, 28, 30
Silver, 49–50
Sklodowska, Marie. *See* Curie, Marie Sklodowska
Sky, 130–31
Sleet. *See* Precipitation
Smog, 135
Snow. *See* Precipitation
Soap, 61
Solar system, 191–93, 212
Solder, 50
Solids, 42, 44–48
Solstices, 184
Solubility, 38
Solutions, 60–61
Sonic booms, 115
Sorting, 27
Sosigenes, 219

Sound, 110–12, 114–18; barrier, 115–16
South Pole. *See* Poles, geographic
Southern Cross, 194
Space, 256–57, 261, 263–65
Specific gravity, 35, 37
Specific heat, 37
Spectra, 64–65, 67
Spectroscopy, 20, 197
Stahl, Georg Ernst, 11
Standard Time, 233–35
Starlight, aberration of, 177
Stars, 123, 157, 173, 193–97
Static electricity, 139–40
Steady-state universe, 199, 264
Steam. *See* Water
Steel, 49, 91
Strassmann, Fritz, 146
Stratosphere, 58
Styrofoam, 46
Sublimation, 42, 102
Sun: appearance of, 130–31, 179, 194; apparent motion of, 157, 164; brightness of, 126; as center of solar system, 167–69, 173–74, 191; influence on tides, 191; nuclear fusion of, 151; radiation of, 131–33, 137, 183; Sunburn, 133 *See also* Astronomy; Eclipses; Day
Sundial, 225

T
Teflon, 47, 149
Telescope, 167, 198
Temperature, 91–92, 94–96
Thales, 3–4, 136, 139, 159–60, 190
Thermodynamics, 106, 205–6
Thermometers, 92
Thermonuclear bomb, 153
Thermos bottle, 109
Thermostats, 92
Thomson, Joseph J., 62, 66, 143
Thomson, William, 96
Thompson, Benjamin, 74–75
Thrust, 86
Thunder, 115, 123
Tides, 163, 173, 190–91
Time, 203–206; end of, 239, 241; relative nature of, 232, 257–58, 262
Time zones, 235–37
Torricelli, Evangelista, 57

Traité Elémentaire de Chemie (Lavoisier), 13
Transparency, 121–22, 131
Tropics, Cancer and Capricorn, 184
Troposphere, 58
Tuning fork, 229
Tycho Brahe, 166
Tyndall, John, 130
Tyndall effect, 130–31

U-V
Ultrasound, 117
Ultraviolet radiation, 132, 135
Uncertainty principle, 269
Unidentified flying objects (UFOs), 123
Uniformitarianism, 208–9
Universal gas laws, 56
Universal gravitation, 170–73, 190–91, 241, 251–52
Universal Standard Time, 236
Universe: ancient theories of, 157, 159–62; current theories of, 197–99, 239; origins of, 210–13, 239; size of, 263–65
Uranium, 62–64, 144–50, 209, 211
Uranium Project, 147–48
Uranus, 158
Urban VIII (Pope), 168–69
Ussher, James, 210

Vaporization, latent heat of, 98–103
Velocity, momentum and, 85–86

Venus, 158, 167–68, 191
Vinegar, 61
Viscosity, 51
Vision, 119–21
Volta, Alessandro, 140–41
Voltage, 141
Volume, 33, 40, 54–57, 92

W-Z
Walton, Ernest, 145
Water, 52–54; composition of, 14–15; density of, 35, 37; freezing and melting of, 53, 97, 101; heat capacity of, 37–38; vaporization and condensation of, 98
Watt, 141
Weather, 37, 58–59, 103–6, 182
Week, 220–21
Weight, 34, 172–73
Weizsacher, Carl Friedrich von, 212
Wetness, 52
Wind, 37, 58–59, 104, 182
Wöhler, Friedrich, 31

X-rays, 62, 135

Yeager, Charles E., 115
Year, 214–16, 225

Zodiac, 196
Zones, climactic, 184